Springer Proceedings in Business and Economics

Springer Proceedings in Business and Economics brings the most current research presented at conferences and workshops to a global readership. The series features volumes (in electronic and print formats) of selected contributions from conferences in all areas of economics, business, management, and finance. In addition to an overall evaluation by the publisher of the topical interest, scientific quality, and timeliness of each volume, each contribution is refereed to standards comparable to those of leading journals, resulting in authoritative contributions to the respective fields. Springer's production and distribution infrastructure ensures rapid publication and wide circulation of the latest developments in the most compelling and promising areas of research today.

The editorial development of volumes may be managed using Springer Nature's innovative EquinOCS, a proven online conference proceedings submission, management and review system. This system is designed to ensure an efficient timeline for your publication, making Springer Proceedings in Business and Economics the premier series to publish your workshop or conference volume.

This book series is indexed in SCOPUS.

Shani D. Carter
Editor

COVID-19, Supply Chain, Climate Change, and Sustainable Development in Africa

Current Business Issues in African Countries (CBIAC) Conference, Staten Island, NY, USA, April 7-8, 2022

Springer

Editor
Shani D. Carter
Nicolais School of Business
Wagner College
Staten Island, NY, USA

ISSN 2198-7246 ISSN 2198-7254 (electronic)
Springer Proceedings in Business and Economics
ISBN 978-3-031-26120-6 ISBN 978-3-031-26121-3 (eBook)
https://doi.org/10.1007/978-3-031-26121-3

This Springer imprint is published by the registered company Springer Nature Switzerland AG
The registered company address is: Gewerbestrasse 11, 6330 Cham, Switzerland

To my father, Edward B. Carter, Sr. whose thirst for knowledge was unquenchable.

Preface

This volume is a valuable addition to the literature on 21st century African business issues. It covers an array of industries across the continent and successfully merges the discussion of theory and practice. The volume's theme is synergies between COVID-19, supply chain, sustainable development, and climate change, and the variety of topics makes clear that all business issues affect each other.

This volume contains an exciting variety of topics demonstrating the wealth of business opportunities across the continent. Within supply chain, the topics examined include the effects of COVID-19 on buying decisions of urban dwellers, sustainable fresh produce, and food contamination. Within sustainable development, topics examined include consumer protection policies, virtual technology, and monetary policy. Within entrepreneurship, topics examined include African-led startups, women entrepreneurs, and religiosity and business owners. Readers will gain insight into theories and tactics at the forefront of growing African businesses.

Current Business Issues in African Countries Conference (CBAIC)

The 1st CBIAC was held in person in New York City in April 2017 (https://www.cbiac.net/). The impetus of the conference is the discussions held at UNCTAD 14 (United Nations Conference on Trade and Development) in Nairobi, Kenya, in July 2016 (https://unctad.org/). The 1st CBIAC gathered dozens of business leaders, NGO leaders, researchers, and students from Nigeria, Kenya, Morocco, and New York City to present information on the current state of business in African countries. Issues discussed revolved around the reciprocal impact of businesses, government, and society. Participants interacted with the Wagner community and participated in campus life.

The 2nd CBIAC was held online in April 2020, during the height of the pandemic, and the online format allowed participation from a broader range of countries. As a result, the 3rd CBIAC was held in a hybrid format in April 2022, where all presentations were in-person in New York and online.

The theme of the 3rd CBIAC in April 2022 was "Supply Chain, Climate Change, COVID-19, Sustainable Development, Entrepreneurship," which mirrored the theme of UNCTAD 15 (United Nations Conference on Trade and Development), which was held in Barbados in October 2021. At the 3rd CBIAC, presenters from Cameroon, Canada, Ghana, Kenya, Liberia, Morocco, the Netherlands, Nigeria, Romania, South Africa, Saudi Arabia, the United Kingdom, and the United States gave 22 presentations on a wide variety of business issues.

We are delighted to present a selection of the research presented at the 3rd CBIAC in this volume.

The 4th CBIAC will be held in Agadir, Morocco, in April 2023 and will again have a hybrid format to ensure a high level of participation globally. The theme of the 4th CBIAC is "Management and Resilience of African Organizations in Times of Crisis," and we look forward to sharing a wealth of information to advance African economies. It is our hope that this volume and CBIAC will expand knowledge of the business climate in African countries and promote sustainable development.

New York, NY, USA Shani D. Carter

Contents

Part I
Introduction

Chapter 1
Synergies Between COVID-19, Supply Chain, Sustainable Development, and Climate Change: Lessons from UNCTAD

Shani D. Carter

1.1 Introduction

The United Nations General Assembly created the United Nations Conference on Trade and Development in 1964 as a permanent intergovernmental body (UNCTAD, 2022a).

The mission of UNCTAD is to support developing countries as they increase their involvement in the global economy. UNCTAD provides analysis and technical assistance and assists with consensus building.

One significant activity of UNCTAD is to offer quadrennial conferences, gathering world leaders to discuss and negotiate contemporary issues. The 2016 gathering (UNCTAD 14) was held in Nairobi, Kenya, and focused on treaties and fair trade. The 2021 gathering (UNCTAD 15) was held in Bridgetown, Barbados, and concentrated on COVID-19, supply chain, climate change, and sustainable development (UNCTAD, 2022b).

1.2 Discussions at UNCTAD 15

1.2.1 Interdependence and Synergies

Synergies between countries and between issues were the guiding theme of UNCTAD 15. To this end, Isabelle Durant, Deputy Secretary-General, UNCTAD, stated, "The interdependence of the whole world must guide our roadmap" (Durant,

S. D. Carter (✉)
Nicolais School of Business, Wagner College, Staten Island, NY, USA
e-mail: shani.carter@wagner.edu

S. D. Carter (ed.), *COVID-19, Supply Chain, Climate Change, and Sustainable Development in Africa*, Springer Proceedings in Business and Economics,
https://doi.org/10.1007/978-3-031-26121-3_1

2021). The COVID-19 pandemic from 2020 to 2021 showed that all aspects of the global economy operate synergistically and that what occurs in one part of the world affects the whole world. For example, the lockdowns governments ordered to decrease the spread of COVID-19 caused a significant reduction in production and transportation, which negatively affected supply chains and reduced the ability of developing countries to maintain their economies. Simultaneously, climate change represents an existential threat to SIDS (small islands and developing states), many of which were hardest hit by the economic effects of COVID-19.

1.3 COVID-19

The COVID-19 pandemic, which began in early 2020, laid bare and exacerbated inequalities within and between countries. The lack of sustainable development, trade inequalities, debt service, and lack of water were problematic for many countries in addressing the pandemic.

Many countries used debt to purchase healthcare supplies for people infected, vaccines, and PPE (personal protective equipment). This COVID-19 debt will hamper future development and push countries back into poverty, making it more difficult for them to hit sustainable development goals.

Further, many small countries lost 50–70% of their revenue due to the pandemic, and countries that rely on tourism were significantly impacted. The lower revenue and decreased tax payments meant these countries had little money to obtain necessary medical supplies.

Small countries with sufficient funds to purchase medical supplies were also treated inequitably. Specifically, large, wealthy countries ordered five times as many vaccine doses as they needed, leaving few vaccine doses available for small countries to purchase. As a result, by October 2021, small countries had vaccinated only 2–5% of their populations, leading to the emergence of COVID-19, the variants affecting the large, wealthy countries.

1.4 Climate Change

There are 18–58 SIDS (small island developing states) depending upon the definition, but it is generally agreed that SIDS are remote, highly vulnerable to shocks (economic and environmental), and cannot obtain economies of scale due to their small geographic size (UNCTAD, 2022c).

Small, developing countries are not the cause of climate change, but they suffer the most. Climate change is an existential threat for many countries, especially the SIDS. If the Paris Accord to limit global warming to 1.5 degrees Celsius, above pre-industrial levels (United Nations, 2022a), is not reached, many SIDS will cease to exist due to rising sea levels.

Climate change, trade, and peace are inextricably linked. Specifically, climate change has caused droughts which negatively impact production. The lack of industry and economic opportunity for populations create conditions that could lead to the destruction of peace. For example, in northeast Nigeria, the level of Lake Chad has decreased by 90% since 1960, and the related loss of fish and industry is thought to have given fertile ground to Boko Haram.

1.5 Sustainable Development

The term sustainable development refers to managing businesses and economies in a manner that promotes prosperity and protects the planet and people (United Nations, 2022b). Engaging in sustainable development requires the recognition of synergies between economic growth and social needs such as health, education, job opportunities, the environment, and climate change.

Due to global trade, there is a lack of substantiable development in developing countries. For example, annually, trillions of dollars in raw materials are exported from developing countries and sent to developed countries. The lack of value-added industries in developing countries makes economic advancement in developing countries unsustainable.

There are opportunities to improve sustainable development in developing countries through increased trade with other developing countries. For example, intra-Africa trade via the African Free Trade Market and increased multilateralism between Africa and the Caribbean have helped African countries improve their economies. The multilateralism within Africa and between Africa and the Caribbean can increase sustainable development and pressure developed countries to have genuine global multilateralism. In addition, there are massive opportunities for FDI (foreign direct investment) into Africa, which would benefit economies globally.

In 2015, the United Nations Member States adopted 17 Sustainable Development Goals (SDGs) along with a 15-year plan to achieve the Goals by 2030. The SDGs are shown in Fig. 1.1.

1.6 Supply Chain

Throughout the late twentieth century and the early twenty-first century, businesses increased multilateral production and created increasingly complicated supply chains. The pandemic uncovered problems in the supply chain in virtually all industries due to lockdowns closing factories and illness, which decreased the number of transportation workers.

There were significant shortages in critical items such as PPE (personal protective equipment), medical supplies, and microchips. Considerable pressure was placed on supply chains, especially for companies with single-source suppliers in

Fig. 1.1 United Nations Sustainable Development Goals. (Source: https://www.un.org/sustainabledevelopment; https://sdgs.un.org/goals)

other countries. For example, microchip shortages in Asia negatively affected automobile production in Europe.

Some companies pivoted to address the shortages. For example, whisky manufacturers began making hand sanitizer, and clothing manufacturers began making PPE. Further, hotels became quarantine centers, etc.

New difficulties are being faced due to the pandemic-related movement away from offshoring production and returning to domestic production. For example, there is increasing nationalism wherein some citizens and government officials would prefer to limit their countries' interdependence with other countries.

The pandemic caused increased shipping costs. These shipping costs are still skyrocketing, raising serious inflationary concerns. This inflation negatively impacts food insecurity in SIDS (small island developing states).

1.7 Entrepreneurship

Entrepreneurship could be an engine to drive sustainable development in developing countries and SIDS. There are, however, many barriers to entrepreneurship in developing countries. For example, the power of large technology companies hinders the work of small businesses and hampers creativity. Trade is now not just at shipping ports but is conducted digitally, and trade is now in private hands (i.e., large technology companies) more than ever. This private oligopoly must be addressed if development is to become sustainable and entrepreneurship is to flourish.

Another barrier to entrepreneurship is taxation. Specifically, there is no global framework for corporate tax for countries to obtain revenue based on the activity that occurs in their countries. This system allows large companies to avoid paying their fair share of taxes globally, representing an unfair disadvantage against entrepreneurs.

1.8 Conclusions

COVID-19 demonstrated that supply chain, climate change, sustainable development, and entrepreneurship are interconnected. Large, developed countries have, at times, promoted treaties, trade, and economic development in ways that do not support sustainable development, and this unfairness should be addressed. Fair sharing of resources must be promoted under the framework of the UN SDG. In the future, all countries should cooperate to ensure fair trade treaties and contracts, fair use of raw materials, and fair access to medical resources, technology, and financing. Last, countries should cooperate to decrease pollution to lessen the effects of climate change.

References

Durant, I. (2021). Remarks by deputy secretary-general, UNCTAD. United Nations conference on trade and development (UNCTAD15), 5 October 5, 2021.

UNCTAD. (2022a). About UNCTAD. United Nations conference on trade and development. https://unctad.org/about

UNCTAD. (2022b). UNCTAD15. United Nations conference on trade and development. https://unctad.org/meeting/fifteenth-session-united-nations-conference-trade-and-development-unctad-15

UNCTAD. (2022c). SIDS in context. UNCTAD. United Nations conference on trade and development. https://dgff2021.unctad.org/

United Nations. (2022a). The Paris agreement. What is the Paris agreement? United Nations. https://unfccc.int/process-and-meetings/the-paris-agreement/the-paris-agreement

United Nations. (2022b). United Nations sustainable development goals. United Nations. https://www.un.org/sustainabledevelopment/

Part II
Supply Chain

Chapter 2
COVID-19: Effects on Buying Decisions and Patterns Among Urban Dwellers in a Sub-Sahara African Market, A Case of Lagos Metropolis

Ayodele Oniku, Olamide R. Okunnu and Owolabi Kuye

2.1 Introduction

The outbreak of the pandemic has multiple effects on global productivity which invariably affected every economy, and the adverse effects are felt across the different sectors (ILO Monitor, 2020; Alayza & Caldwell, 2021; Sharifi & Khavarian-Garmsir, 2020). In fact, the world is currently confronted with one of the biggest public health risks in living memory. As many parts of the world continue to struggle with the pandemic, and the scientific community and nations scramble for vaccines, there is ample evidence to show that COVID-19 is affecting a broad range of sectors, including international economic and trade activities, social and cultural relationships, healthcare, and food security. COVID-19 has rapidly disrupted what is considered as a "normal" way of life (Alayza & Caldwell, 2021).

With an estimated 90% of all reported COVID-19 cases (UN-Habitat), urban areas have become the epicenter of the pandemic. The size of their populations and their high level of global and local interconnectivity make them particularly vulnerable to the spread of the virus. Cities are home to most of the world population and are centers of economic growth and innovation. However, the high concentration of people and activities in cities make them vulnerable to various stressors such as

A. Oniku (✉)
University of Lagos, Lagos, Nigeria
e-mail: ooniku@unilag.edu.ng

O. R. Okunnu (✉)
Connect Marketing Services, Lagos State, Nigeria

O. Kuye
Department of Business Administration, Faculty of Management Sciences,
University of Lagos, Akoka-Yaba, Nigeria
e-mail: okuye@unilag.edu.ng

S. D. Carter (ed.), *COVID-19, Supply Chain, Climate Change, and Sustainable Development in Africa*, Springer Proceedings in Business and Economics,
https://doi.org/10.1007/978-3-031-26121-3_2

natural and man-made disasters. Understanding this, over the past few decades, a vast body of research has been published on the impacts of a wide range of disasters on cities, and necessary planning, recovery, and adaptation measures that need to be taken to deal with those disasters (Sharifi & Khavarian-Garmsir, 2020). Studies have shown that the COVID-19 health crisis has expanded to a crisis of urban access, urban equity, urban finance, safety, joblessness, public services, infrastructure, and transport, all of which are disproportionally affecting the most vulnerable in society. Estimates from the World Bank and UN entities suggest that local governments may on average lose 15% to 25% in revenues in 2021 (Savarimuthu et al., 2020).

Thus, the study focuses on identifying the effects of COVID-19 pandemic on the living, lifestyle, and experiences of urban dwellers, and specifically the middle-class' ladder of the social class scale. The choice of the economic class in the society is primarily based on the class ability and propensity to spend on arrays of goods and services to meet their basic and social needs. Importantly, the middle class in any society is regarded as the oil of spending that keeps the engine of every economy functional.

2.2 Literature Review

As a result of COVID-19 pandemic, people significantly changed purchasing behaviors (Dooley et al., 2010; Foxall, 1979; Lamming, 2000; Habel et al., 2020). Most have adapted to online and home delivery purchases by the social factor, local consumption, and consumers' attitude.

Although consumption develops habitually and contextually, recent studies on consumer behaviors describe that purchasing habits change from different contexts (Jagdish, 2020). These studies include the social context, technological context, context of coexistence, and external context. Thus, the multidimensional changes in purchasing habits have impactful markets not only on the consumers but equally the adaptation of marketing mechanisms like retailing, supply chain, production, etc. to ensure that consumer retention and satisfaction are achieved. The importance of understanding and effectively managing the dimensions is more desirable in recent time because of the unexpected changes in purchase behaviors and patterns as a result of COID-19 pandemic. Consumers, especially the middle class, are becoming more rational due to irregularity in income flow, and health-conscious because of the rise in health budget occasioned by the pandemic.

The effects of COVID-19 on the middle-class urban dwellers' buying decision are expressed in the following ways:

(i) Hoarding: Hoarding is a reaction to managing uncertainty and fear of the future supply of products for basic needs. Hoarding is a common practice in war situations, hyperinflation, and the black market. In addition to hoarding, the gray market is also emerging where unauthorized intermediaries hoard the product and raise prices.

(ii) Improvisation: Improvisation to manage the shortage of products or services is an innovative practice. It is an alternative consumption pattern that can lead to improvisational systemic consumption behavior in crisis situations (civil wars, natural disasters, authoritarian regimes, and pandemics).
(iii) Accumulated demand: In times of crisis and uncertainty, the general trend is to postpone the purchase and consumption of nonessential products or services. The main consequences of denying leisure and recreation services are the pandemic period's effects since superfluous products or goods find a way to be marketed.
(iv) Embracing digital technology: Consumers have adopted new technologies and their applications. They have drastically changed the nature and scope of word-of-mouth advice and recommendations, and information exchange in a highly compressed cycle such as the pandemic.
(v) Adjustments to the working day: In the pandemic's face and its consequences, the boundaries between works/businesses and home and between tasks, chatting, and time spent in family life have blurred. It is necessary to respect work schedules and align them to the children's school activities, as they used to be before the pandemic, but now with technological mediation, the need to make the home more efficient and effective is incumbent on the adjustments to working hours and days.
(vi) School meetings with friends and family: It seems that school reunions (from primary education to university) and family weddings are now regular and scheduled gatherings under the same approach to share information and experiences from a remote link.
(vii) Discovery of talent: Consumers are becoming producers of their products or services. In other words, in the pandemic's face, the consumer they carry within, and even the entrepreneur, has emerged, given the health circumstances and market access. They are discovering the potential for innovation and commercial success.

Largely, the change in purchase behaviors and habits among the middle class is measured from the areas discussed below:

2.2.1 Adoption of Online Purchase Option

The resultant effect of the COVID-19 pandemic has led to the compression of 10 years of e-commerce into few months, as several consumers now rely more on e-commerce for almost all purchases. The reason been adduced is the widespread stay-at-home order and concerns for the rapid transmission rate of the virus. And this drift to adoption of online purchase option mindset did not just escalate in countries where online shopping was already widely accepted but also evident in cultures where in-person, local, cash-reliant, and daily shopping was the erstwhile norm.

2.2.2 Health and Pharmaceutical Budget

At the inception of the COVID-19, the changes the pandemic would elicit in the health and pharmaceuticals budgets were uncertain. Though many anticipated an attendant increase in such spending, however, there were other variables to also be taken into consideration.

Hence, with the incorporation of all variables, it was noted that healthcare use and spending dropped headlong, as many opted out of non-severe checks in order to increase hospital capacity or to observe social distancing measures to curb the spread of the virus. However, though telemedicine and off-the-shelf purchases of pharmaceuticals recorded an unprecedented spike, it did not suffice to augment for the drop in in-person care. Nevertheless, expenditure in healthcare and pharmaceuticals regained momentum as the year progressed, and spending began to rebound as in-person care resumed for hospital and lab services and COVID-19 testing became more widely available. The decrease in health spending in 2020 especially can be said to be a reflection of a decrease in utilization for non-COVID medical care.

2.2.3 Vegetable and Food Budget

Vegetables are plants or parts of plant used for food. In response to the outbreak of the COVID-19 pandemic, most households took a drastic turn from visiting restaurants for their food consumption to homemade meals as stay-at-home orders went into effect. This resulted in demand shocks and problems with supply chains that culminated to increased volatility in import, export, producer, and consumer prices in the months following the onset of the COVID-19 pandemic. The outcome was a spike in the prices of food and vegetables due to the shifting economy brought on by the pandemic.

2.2.4 Holidays and Tourism Budget

There was a drastic reduction witnessed in spending in this segment as the sit-at-home order and travel restrictions impacted heavily on the sector. Hence, many people were forced to either postpone or cancel their trips.

Pandemics like COVID-19 result in a disruption in the lifestyle and buying pattern of a consumer and adversely impact the global economy. Meghan Verma in her article – Covid-19 Impact on Buying Behaviour – present study results which reveal that during pandemics, consumers have realized the importance of hygiene products, environment-friendly products, regional (local) products, and satisfaction beyond shopping; these factors determine their willingness to buy locally made products. The behavioral responses during disease outbreaks such as Ebola, SARS, MERS, swine flu, and dengue have been studied in the past (Balinska & Rizzo, 2009). Such changes were noted during pandemics and outbreaks such as SARS because of motivated individuals and government policies (Wen et al., 2005). The areas affected by pandemics generally witness unemployment, uncertainties, and economic recession. Individuals take safety measures to reduce the perceived risk during pandemic situations due to ambiguity and unpredictability (Brug et al., 2009). An increase in the purchase of food, face masks, and sanitizers was reported during the outbreak of swine flu (Goodwin et al., 2009).

Furthermore, Sharma and Tikka (2020) document that during pandemics, the concept of consumer ethnocentrism tends to thrive. It is based on three views: first, the consumer's fear of economically harming his/her own country by buying foreign products; second, the morality in buying imported products; and, third, a personal prejudice against imports. Ethnocentric consumers believe that purchasing imported products is unpatriotic, causes loss of jobs, and hurts the domestic economy. Consumer ethnocentrism plays a significant role in the purchase of domestic brands but has a moderate effect on consumer purchase intentions for foreign brands. Product advertisements that display ethnocentrism may not encourage the purchase of domestic brands but can discourage foreign brand purchases (Han & Guo, 2018). Also, Peck and Childers (2006) found that at the point of purchase, signs, displays, and packaging encourage a touch of the product, which play a crucial role in impulse purchase and the availability of money. During pandemic, impulse buying could be seen in panic buying, resulting in stockpiling of goods including food (Wang & Na, 2020).

Abe (2020) in her report on "Market Trends and D2C Opportunities in the COVID-19" observed that during the pandemic, people are spending less of their income on items perceived as nice-to-have or nonessential (such as clothing, shoes, makeup, jewelry, games, and electronics). The report also stated edible products are expected to have an increased demand and nonedible products shall have a moderate need globally, thereby decreasing demand which includes homecare, cosmetics, and personal care products. The method of research applied was through an analysis of relevant published surveys during COVID-19 and an in-depth discussion with few marketing professionals and consumers. Further literature was also explored to find out the reasons of these behavioral changes in consumer decision-making process in times of crisis (Kantar, 2020).

2.3 Methods

The population of study is all residents of Lagos metropolis. The choice of the population is because they were deemed appropriate to elicit relevant information useful for the investigation. The target population figure is 14.8 million according to

UN World Urbanization prospects, 2018. The population was adopted because there is no verifiable data about middle-class population for the metropolis; however, the survey exercise was targeted at the middle class in the society based on residential cluster, work positions/designations, and business ownership. The study adopted quantitative approach, and the questionnaires were administered online via Google link. Multiple regression analysis was used to interpret the data. With the application of Taro Yamanne model, the total sample size for the study was 420, and a total of 400 (95.2%) questionnaires were successfully completed and returned. The Cronbach alpha coefficient for the reliability and internal consistency test showed a range of 0.6–0.80.

2.3.1 The Hypotheses Are Listed Below

Ho: COVID-19 and the resultant lockdown do not have any effect on online purchase behaviors among urban dwellers in Lagos metropolis.
Ho: COVID-19 and the resultant lockdown do not have any effect on health budget and pharmaceutical needs of urban dwellers in Lagos metropolis.
Ho: COVID-19 and the resultant lockdown do not have any effect on vegetable budget and foods bill among urban dwellers in Lagos metropolis.
Ho: COVID-19 and the resultant lockdown do not have any effect on household budget on holiday trips among urban dwellers in Lagos metropolis.

2.4 Data Analysis and Findings

2.4.1 Test of Hypotheses

The examination of the relationship in existence between the variables identified in the study as stated in the research objectives, research questions, and the hypothesis would be analyzed under this section using multiple regression analysis.

2.4.1.1 Decision Criteria

The analysis below was extracted from the SPSS output file outlining the adjusted coefficient of determination, adjusted R2, and the correlation coefficient (R). In addition, the ANOVA table showing the F statistic for assessing the claim that there is significant relationship between independent and dependent variables. The significant value is the P value. Thus, we will accept the claim that there is significant relationship between your independent and dependent variables if $p < \alpha$.

In the last column of the coefficient box, the P value for individual t-test for our independent variable is given. It claims that there is a relationship between the independent variable (in the corresponding row) and the dependent variable. Thus, we will accept the claim that there is significant relationship between the independent variable (in the corresponding row) and dependent variable if $p < \alpha$.

NB: alpha (α) at 5% significant level is 0.05.

2.5 Multiple Regression Analysis

Model Summary[a]

Model	R	R square	Adjusted R square	Std. error of the estimate	Durbin-Watson
1	.341[b]	0.117	0.108	0.59924	1.837

[a]Predictors (constant): COVID-19 (online purchase)
[b](The Dependent: UDBD – holiday/tourism, health/pharmaceutical, household/health, vegetable/food/bills)

ANOVA[a]

Model		Sum of squares	df	Mean square	F	Sig.
1	Regression	18.708	4	4.677	13.024	.000[b]
	Residual	141.840	395	0.359		
	Total	160.548	399			

[a]Dependent: UDBD – (holiday/tourism, health_pharmaceutical, household/health, vegetable_food_bills)
[b]Predictors (constant): COVID-19 (online purchase)

Coefficients[a]

Model		Unstandardized coefficients		Standardized coefficients		
		B	Std. error	Beta	T	Sig.
1	Constant	1.618	0.178		9.087	0.000
	Health_pharmaceutical	0.094	0.056	0.092	1.675	0.095
	Vegetable_food_bills	0.105	0.055	0.112	1.920	0.056
	Household/health	0.223	0.054	0.232	4.156	0.000
	Holiday/tourism	−0.020	0.057	−0.020	−0.358	0.721

[a]Independent variable: COVID-19 (online purchase)

2.6 Interpretation of Results

Hypothesis 1
H0: COVID-19 and the resultant lockdown do not have any effect on health budget and pharmaceutical products in Lagos metropolis.

The ANOVA result further reveals that the relationship is not significant as the P value (0.095) is more than the beta value (0.05). This is a confirmation of the regression result that the online purchases in Lagos metropolis has had no effect on the health/pharmaceuticals.

Decision
Since P (0.095) > β (0.05), H0 is accepted and H1 is rejected, meaning that the relationship between the online purchases and health and pharmaceuticals is not statistically significant both not at 1% and 5% level of significance.

Since for hypothesis 1 the significance is 0.470 which is far more than 0.05, the null hypothesis (H0) is accepted, and the alternate hypothesis (H1) is rejected and concludes that COVID-19 and the resultant lockdown do not have any effect on online purchases for health and pharmaceuticals of urban dwellers middle class' buying decision in Lagos metropolis.

Hypothesis 2
Ho: COVID-19 and the resultant lockdown do not have any effect on vegetable budget and food bills in Lagos metropolis.

The ANOVA result further reveals that the relationship is not significant as the P value (0.056) is more than the beta value (0.05). This is a confirmation of the regression result that the online purchases resulting from the COVIS-19 pandemic in Lagos metropolis has had no effect on the vegetable budget and food bills of urban dwellers middle class' buying decision.

Decision
Since P (0.056) > β (0.05), H0 is accepted and H1 is rejected, meaning that the relationship arising from online purchases due to COVID-19 and the resultant lockdown and vegetables and food bills is not statistically significant both not at 1% and 5% level of significance.

Since for hypothesis 2 the significance is 0.056 which is far more than 0.05, the null hypothesis (H0) is accepted, and the alternate hypothesis (H1) is rejected and concludes that COVID-19 and the resultant lockdown do not have any effect on vegetable budget and food bills of urban dwellers middle class' buying decision in Lagos metropolis.

Hypothesis 3
H0: COVID-19 and the resultant lockdown do not have any effect on health for households in Nigeria.

The ANOVA result further reveals that the relationship is significant as the P value (0.000) is less than the beta value (0.05). This is a confirmation of the regression result that the online purchases resulting from the COVID-19 pandemic in Lagos metropolis has a high, positive, and significant effect on household/health of urban dwellers middle class' buying decision.

Decision
Since P (0.000) < β (0.05), H1 is accepted and H0 is rejected, meaning that COVID-19 and the resultant lockdown and household/health are statistically significant both not at 1% and 5% level of significance.

Since for hypothesis 3 the significance is 0.000 which is less than 0.05, the null hypothesis (H0) is rejected, and the alternate hypothesis (H1) is accepted and concludes that COVID-19 and the resultant lockdown has high, positive, and significant effect on household/health of urban dwellers middle class' buying decision in Lagos metropolis.

Hypothesis 4
H0: COVID-19 and the resultant lockdown do not have any effect on the budget on holiday and tourism in Lagos metropolis.

The ANOVA result further reveals that the relationship is not significant as the P value (0.721) is greater than the beta value (0.05). This is a confirmation of the regression result that the COVID-19 and the resultant lockdown has no effect on holiday/tourism of urban dwellers' buying decision.

Decision
Since P (0.721) > β (0.05), H0 is accepted and H1 is rejected, meaning that COVID-19 and the resultant lockdown and household/health are not statistically significant both not at 1% and 5% level of significance.

Since for hypothesis 4 the significance is 0.721 which is greater than 0.05, the null hypothesis (H0) is accepted, and the alternate hypothesis (H1) is rejected and concludes that COVID-19 and the resultant lockdown do not have any effect on holiday/tourism of urban dwellers in Lagos metropolis.

2.7 Discussion of Findings and Conclusion

The position of the study supports the findings of Fatoye et al. (2021), Hesham et al. (2021), Islam et al. (2020), Prentice et al. (2021), Abdulkareem et al. (2020), which noted that lockdown and total adherence to non-pharmaceutical protocols assisted in reducing the impact of COVID-19 pandemic. This is especially significant as the threat of fourth wave ravages most European countries leading to some of them already resorting back to lockdown, despite appreciable vaccination rate. Worthy of note is Romania and Germany, which address the outbreak as the disease of the unvaccinated. Thus, the urban dwellers see the need to spend more on health and pharmaceutical-related needs more important to remain healthy and COVID-19-free. However, budgets on traveling, vegetable, and patronage of online purchase were not seriously affected during the period among the middle-class urban dwellers.

References

Abdulkareem, S., Augustijn, E., Filatova, T., Musial, K., & Mustafa, Y. (2020). Risk perception and behavioural change during epidemics: Comparing models of individual and collective learning. *PLoS One, 15*(1), e0226483. https://doi.org/10.1371/journal.pone.0226483

Abe, S. (2020). Market trends and D2C opportunities in the COVID-19 landscape. The AdRoll Blog. https://www.adroll.com/blog/marketing/market-trends-and-d2c-opportunities-inthecovid-19-landscape. Accessed Mar 2022.

Alayza, N. & Caldwell, M. (2021). Financing climate action and the COVID-19 pandemic: An analysis pf 17 developing countries. World resources institute (Working Paper), Nov, 1–32.

Balinska, M., & Rizzo, C. (2009). Behavioural responses to influenza pandemics: What do we know? *PLoS Curr, 1*. https://doi.org/10.1371/currents.rrn1037. Accessed Apr 2022.

Brug, J., Aro, A. & Richardus, J. (2009). Risk Perception and Behaviour towards Pandemic Control of Emerging Infectious Diseases, International *Journal of Behavioural Medicine, 16*(1), 3–6.

Dooley, K. J., Yan, T., Mohan, S., & Gopalakrishnan, M. (2010). Inventory management and the bullwhip effect during the 2007–2009 recession: Evidence from the manufacturing sector. *Journal of Supply Chain Management, 46*(1), 12–18.

Fatoye, F., Gebrye, T., Arije, O., Fatoye, C., Onigbinde, O., & Mbada, C. (2021). PRS17 economic impact of COVID-19 lockdown on households in Nigeria. *Value in Health, 24*, S215–S216. https://doi.org/10.1016/j.jval.2021.04.1081

Foxall. (1979). Industrial buying during recession: Farmers' tractor purchases, 1977–78. *Management Decision, 17*(4), 317–325.

Goodwin, R., Haque, S., Neto, F., & Myers, L. (2009). Initial psychological responses to Influenza A, H1N1 ("Swine flu"). *BMC Infectious Diseases, 9*(1). https://doi.org/10.1186/1471-2334-9-166

Habel, J., Jarotschkin, V., Schmitz, B., Eggert, A., & Plotner, O. (2020). Industrial buying during the Corona virus pandemic: A cross-cultural study. *Industrial Marketing Management, 88*(July), 195–205.

Han, C.-M., & Guo, C. (2018). How consumer ethnocentrism (CET), ethnocentric marketing, and consumer individualism affect ethnocentric behaviour in China. *Journal of Global Marketing, 31*(5), 324–338.

Hesham, F., Riadh, H., & Sihem, N. (2021). What have we learned about the effects of the COVID-19 pandemic on consumer behaviour? *Sustainability, 13*(8), 4304. https://doi.org/10.3390/su13084304

ILO Monitor: COVID-19 and the World of Work. http://hdl.voced.edu.au/10707/533608. Accessed Mar 2022.

Islam, T., Pitafi, A. H., Arya, V., Wang, Y., Akhtarf, N., Mubarik, S., & Xiaobei, L. (2020). Panic buying in the COVID-19 pandemic: A multi-country examination. *Journal of Retailing and Consumer Services, 59*, 1–13.

Jagdish, N. (2020). *Sheth the Howard-Sheth theory of buyer behaviour*. Wiley & Sons, New Delhi, India.

Kantar. (2020). Covid-19 barometer. https://www.kantar.com/Campaigns/Covid-19-Barometer. Accessed Mar 2022.

Lamming, R. (2000). Japanese supply chain relationships in recession. *Long Range Planning, 33*(6), 757–778.

Peck, J., & Childers, T. (2006). If I touch it I have to have it: Individual and environmental influences on impulse purchasing. *Journal of Business Research, 59*(6), 765–769.

Prentice, C., Quach, S., & Thaichon, P. (2021). Antecedents and consequences of panic buying: The case of COVID-19. *International Journal Of Consumer Studies, 46*, 132. https://doi.org/10.1111/ijcs.12649

Savarimuthu, R., Dharshini, M., Kumar, U., Gnanaprakash, A., Lakshmi, S., & Cherian, C. (2020). Assessment of aggressiveness amongst young adults during COVID-19 lockdown, Tamil Nadu, India. *Indian Journal of Continuing Nursing Education, 21*(2), 206. https://doi.org/10.4103/ijcn.ijcn_88_20

Sharifi, A., & Khavarian-Garmsir, A. (2020). The COVID-19 pandemic: Impacts on cities and major lessons for urban planning, design, and management. *Science of the Total Environment, 749*, 142391. https://doi.org/10.1016/j.scitotenv.2020.142391

Sharma, R., & Tikka, S. (2020). COVID-19 online surveys need to follow standards and guidelines: Comment on "Does COVID-19 pandemic affect sexual behaviour? A cross-sectional, cross-national online survey" and "Binge watching behaviour during COVID 19 pandemic: A cross-sectional, cross-national online survey". *Psychiatry Research, 290*, 113173. https://doi.org/10.1016/j.psychres.2020.113173

Wang, H. H., & Na, H. (2020). Panic buying? Food hoarding during the pandemic period with city lockdown. *Journal of Integrative Agriculture, 19*(12), 2916–2925.

Wen, Z., Huimin, G., & Kavanaugh, R. R. (2005). The impacts of SARS on the consumer behaviour of Chinese domestic tourists. *Current Issues in Tourism, 8*(1), 22–38.

Chapter 3
Developing Sustainable Fresh Produce Agriculture Value Chains: Case Study of African Indigenous Vegetables (AIVs) Value Chain in Uasin Gishu County, Kenya

Cynthia Onyangore

Acronyms

AIVs:	African Indigenous Vegetables
CFSP:	County Fiscal Strategy Paper
CIPD:	County Integrated Development Plan
GAHP:	Good Agriculture and Handling Practices
JKUAT:	Jomo Kenyatta University of Agriculture and Technology
KCIC:	Kenya Climate Innovation Center
KEPHIS:	Kenya Plant Health Inspectorate Service
KIRDI:	Kenya Industrial Research and Development Institute
NOREB:	North Rift Economic Bloc
NPI:	Natural Products Industry
TUC:	Trade Up Commodities Ltd.
UoE:	University of Eldoret
VC:	Value Chain
UG:	Uasin Gishu County

3.1 Introduction

Presently, smallholder African Indigenous Vegetable (AIV) producers in Uasin Gishu County must find their own markets for their produce. They will majorly rely on intermediaries for quick sales albeit at unfavorable prices. The other option is to

C. Onyangore (✉)
Trade Up Commodities Ltd., Eldoret, Kenya

Van Hall Larenstein University of Applied Sciences, Velp, The Netherlands

S. D. Carter (ed.), *COVID-19, Supply Chain, Climate Change, and Sustainable Development in Africa*, Springer Proceedings in Business and Economics,
https://doi.org/10.1007/978-3-031-26121-3_3

hold on their produce and sell it directly at local seasonal markets, a process that has seen their produce rejected due to deterioration in quality. This is due to lack of access to the dynamics of markets, improper storage, lack of understanding of the market value chains, and other support required to ensure they are fully integrated into the market processes. These issues breed an environment that is not geared to ensuring inclusivity of small-scale holders in the fresh vegetables value chain processes, and specifically for AIVs. This is the premise on which this analysis was anchored.

The goal of this analysis was to analyze the African Indigenous Vegetable (AIV) value chain in Uasin Gishu County, Kenya. The goal was to define and understand the AIV value chain (stakeholders, their interactions, processes, internal and external environmental influencers, qualitative and quantitative analysis) and identify the bottlenecks in the value chain. This will lead to development of solutions that will ensure a more dynamic, inclusive, and sustainable AIV value chain in Uasin Gishu County.

3.1.1 Reason for Selection

Kebede and Bokelman (2017) observe that even though African Indigenous Vegetables (AIVs) have always been a staple food crop, their promotion in terms of production and marketing is a recent phenomenon. This is because they used to be primarily considered as food for rural and poor households only. They also note that their consumption has increased, and so has demand for them. Abukutsa (2021), a leading researcher in AIVs in Kenya, argues that the increase in the uptake of AIVs will boost immunity, thus keeping lifestyle diseases like obesity at bay. Modi (2021), an agronomist and advocate for underutilized, indigenous, or traditional crops, emphasizes the need for science and education systems to support small-scale farmers producing the crops and improve their lives.

3.1.2 Sources of Information

Desk study and primary research was used to collect both primary and secondary data. Primary information was obtained from interviewing small-scale female processors of AIVs in Uasin Gishu County.

Desk research was conducted using existing data, e.g., financial data, trading reports, and county government reports. Review of existing literature on AIVs was done at Google Scholar, Greeni.nl, and ResearchGate. Boolean operators used in the search were "AND" and "OR." Keywords used included African Indigenous Vegetables, AIVs, Kenya, Uasin Gishu, Smallholders, Value Chain, Governance, and Quality. The date parameters were set for 2017–2021 to generate the most up-to-date information. Books on quality management were also referenced.

3.1.3 Commodity and Study Area

3.1.3.1 African Indigenous Vegetables (AIVs)

Otieno et al. (2019) note that AIVs are vegetable crops whose natural habitat originated in Africa. Rampa et al. (2020) describe AIVS as plants whose leaves, fruits, or roots are acceptable and used as vegetables by rural and urban communities through custom, habit, and tradition. Otieno et al. (2019) further note that the most popular AIVs include both wild and cultivated leafy greens such as pumpkin leaves (*Cucurbita pepo*), amaranth (*Amaranthus* spp.), nightshade (*Solanum* spp.), spider plant (*Cleome gynandra*), cowpea (*Vigna unguiculata*), and jute mallow (*Corchorus olitorius*). They further note that their cultivation is more popular with smallholders because they require fewer inputs and are better adapted to local agroecological conditions. From my observation, smallholder producers of AIVs in Uasin Gishu County usually produce more than one variety of AIV to ensure they are meeting demands of different customers, and this is my justification for analyzing them collectively, instead of picking just one AIV.

3.1.4 Study Area: Uasin Gishu County

3.1.4.1 Devolution of Government

According to the World Bank (2021), the Public Finance Management Act was enacted in 2012 after the decentralization of the Kenya Government. With this decentralization, this meant that each of the 47 county governments had to identify, focus, and address the issues of the local people. As detailed by Uasin Gishu County (2021), the first County Fiscal Strategy Paper (CFSP) was developed, which took cognizance of the economic, geographical, and social positioning of the County first as an epicenter of agricultural productivity in Kenya, second, as a transit point into Western Kenya and East African Region, and third, as a highly cosmopolitan region in which the county government can invest in, build, and leverage on the social diversity it represents. The core pillar of all this was in the investment in agricultural transformation.

3.1.4.2 County Description

The County is located in the North Rift region of Kenya. It covers a total area of 3345.2 Km^2. The County has partnered with seven other counties in the region, namely, Elgeyo-Marakwet, Nandi, Transnzoia, Baringo, Turkana, Samburu, and West Pokot to form an economic bloc referred to as the North Rift Economic Bloc (NOREB) which intends to bring long-term social and economic benefits to the residents of the region. The main economic activity in the County is agriculture both

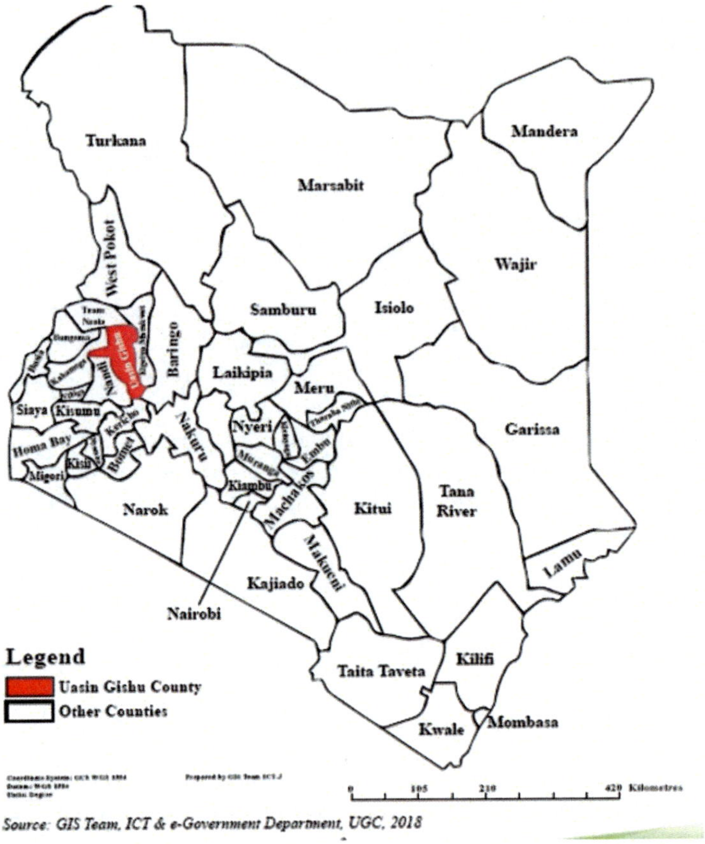

Fig. 3.1 Location of Uasin Gishu County

crop and livestock farming attributed to the County's rich fertile soils and "favorable climatic conditions" (Uasin Gishu County, CIPD, 2018, p. 2). The County has its headquarters in Eldoret Town. According to County Track (2021), the population of UG County at the 2019 population census was 1163, 186, and 80% of the rural population depend on agriculture for household income and food security. Uasin Gishu County (2018) notes that total acreage under food and cash crops is 138, 609 hectares with 132.8 hectares of AIVs. 2 ha–10 ha is considered small-scale farming, and 50 ha and above is considered large-scale farming (Fig. 3.1).

3.1.4.3 Chain Context and External Factors of Influence

The AIV value chain in Uasin Gishu is characterized by highly individualistic behavior among the various stakeholders. There is information asymmetry in the value chain in issues pertaining to pricing of the AIVs, transportation costs, delivery times, and units of measure. For example, purchase from farmers will be in "bunches" or in unweighted sacks yet selling will be in kilograms.

Seasonality of the AIVs is a crucial factor in the pricing and quality decisions. 80% of the farmers depend on the rains for agriculture, with the 20% being the smallholder farmers in the urban areas closer to Eldoret Town, who may have irrigation systems on their farms. During the rainy season, April to September, there is surplus production of AIVs, leading to lower prices and postharvest losses. During the dry season, October to March, there is little production from rural smallholders depending on rain. This leads to less supply in the market and higher prices for the AIVs. Due to climate change, the weather patterns are no longer predictable leading to uncertainty in production.

Consumer preferences are also changing. Eldoret is a cosmopolitan town, with a growing middle class that demands better quality produce and is willing to pay more for that quality. There is a rapid expansion of commercial business spaces that house high-end supermarkets, e.g., Chandarana and specialty green grocers, e.g., *Mboga na Kadhalika* (which translates to vegetables and more). Inasmuch as these provide more avenues for marketing AIVS, they are still inaccessible to smallholders due to the volumes and quality standards required that smallholders are not able to comply with (Fig. 3.2).

Socially, the uptake in AIVs is rapidly becoming a movement! Alulu et al. (2019) argue that AIVs are rich in vitamins and minerals and therefore form an important part of a nutritious diet. They further note that AIVs have shorter growing cycles compared to other crops like maize and maximize on soil nutrients meaning smallholders can earn quicker and a higher return than from other crops. Ngenoh et al. (2019) also assert that AIVs have medicinal value, while noting that AIV producers still face challenges in production and marketing.

Whereas the County of Uasin Gishu has made strides in providing a platform for vegetable sellers in the County by renovating the Eldoret Market and building new

Fig. 3.2 Inside *Mboga na Kadhalika* Greengrocer – Rupa's Mall, Eldoret

Fig. 3.3 Inspection of the Kimumu Market by the Uasin Gishu Governor – H. E Jackson Mandago

market stalls at Ainaptich, Kimumu, Huruma, and Kuinet, they remain largely unoccupied due to the fees traders must pay each month. There is still preference for seasonal markets at these locations, where traders pay Ksh.100 for a day pass to sell the produce they have available (Fig. 3.3).

The overall picture that emerges is one that presents a high potential for the AIV value chain in Uasin Gishu to grow and be more coordinated, a process that will require a multi-stakeholder participatory approach. 80% of producers of AIVs being rural smallholder means there are opportunities to allow them to access the growing markets more effectively and efficiently.

3.1.5 Benchmarking

Mili (2017) defines benchmarking as a direct comparison for selected factors pointing to defined characteristics of a technical or socioeconomic process. The factors to be benchmarked are assessed and compared with similar ones of competing systems or systems considered to be on a stage worthy to be achieved. He outlines the following benchmarking cycle (Fig. 3.4).

For our value chain analysis, we benchmark with Vihiga County. Council of Governors (2021) notes that Vihiga County has implemented commercialization of African Indigenous Vegetable (AIV) ventures as part of a broader agenda under the Vision 2030 flagship. The partners to this project are noted to be (NPI) natural products industry (in the State Department of Culture and Heritage) – provide certified seeds and source for markets, Jomo Kenyatta University of Agriculture and Technology (JKUAT), technical assistance and seed bulking and the Vihiga County

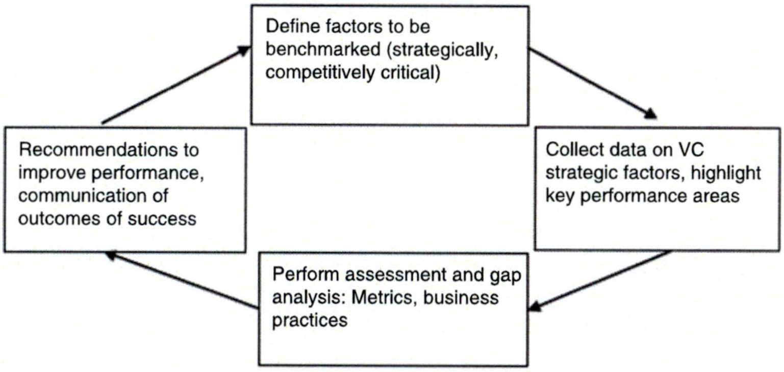

Fig. 3.4 Benchmarking cycle. (Source: Samir Mili, 2017)

Table 3.1 Comparing AIV value chain in Uasin Gishu and Vihiga Counties

	Vihiga County	Uasin Gishu County
Robustness (e.g., chain relations)	Coordination at various levels in the VC	Uncoordination among the various stakeholders
Reliability (e.g., institutional governance like PPPs – public-private partnerships)	Partnerships with farmers, natural products industry, JKUAT and Vihiga Country trading department	Limited partnerships between farmers and private sector companies
Resilience (in innovation systems, e.g., knowledge exchange, technical support, capacity development)	Technical support from JKUAT, NPI, and the County Government of Vihiga	Limited and where present, is part of embedded services from private sector companies

Source: Author's own composition

Department of Trade, and train farmers on AIV production and Good Agriculture and Handling Practices (GAHP) (Table 3.1).

From this comparison, we can deduce that Uasin Gishu County needs to expend more efforts in building an AIV value chain that will be beneficial to stakeholders.

3.1.6 Stakeholder Matrix

Table 3.2 shows the various stakeholders in the AIV value chain in Uasin Gishu County and their roles and functions. It includes the chain actors, chain supporters, and chain influencers. According to Kenya County Guide (2021), Eldoret, Burnt Forest, Matunda, Turbo, Kesses, Kapseret, and Moiben are considered urban areas. The rest of the 90 sublocations are considered rural.

Table 3.2 Value chain actors in Uasin Gishu County AIV value chain

VC level and actor (s)	Role/function
Input supply: Agrovets Seed suppliers	Provision and supply of AIV seedlings and seeds and other inputs required for production
Production: Small-scale rural farmers Small-scale urban farmers Supermarket farms Contract farmers	Production of the African Indigenous Vegetables
Trading: Middlemen Private sector companies Wholesalers	Buying and selling of the AIVs. Intermediaries buy from rural smallholders unable to bring their produce to markets and at low prices and sell to wholesalers. Private companies buy from both rural and smallholders and sometimes from wholesalers to cover shortfall in supply
Processing: Small-scale processors Commercial processers, e.g., Mace Foods	Processing of the AIVs by sorting, washing, drying, fermenting, or boiling and packaging for sale to supermarkets and high-end consumers
Retailing: Local market retailers High-end green grocers Supermarkets	Selling of the AIVs to the end users
Consuming: Low-income consumers High-income consumers Export market (only for high-end processors)	They are the final consumers of the AIVs

3.1.6.1 Chain Actors

These are the stakeholders directly involved in the value chain, and who own the product/commodity/service. They include the following.

3.1.6.2 Chain Supporters

These are the stakeholders that provide support services to the chain actors, e.g., financial services and technical support (Table 3.3).

Table 3.3 Value chain supporters in Uasin Gishu County AIV value chain

Chain supporter	Role/function
Uasin Gishu County	Provides limited AIV seedlings to farmers in rural areas and overseas business services and functions in the County
University of Eldoret (UoE)	Research and training for value addition to the AIVs, e.g., through fermenting and drying
Financial institutions, e.g., Equity Bank	Provision of loans to farmers and other actors in the VC
KIRDI (Kenya Industrial Research and Development Institute)	Provides some technical training to few selected urban smallholders on value addition for AIVs, e.g., packaging

Table 3.4 Value chain influencers in Uasin Gishu County AIV value chain

Influencer	Role/function
Uasin Gishu County	Crafts the policy framework that informs agriculture policy in the county
Kenya Climate Innovation Center (KCIC)	Funding for the AgriBiz program that targets select women and youth. The projects are usually funded by the EU, DANIDA, and World Bank

3.1.6.3 Chain Influencers

Chain influencers provide a (dis)enabling environment for the value chain, e.g., in provision of funding, laws, rules, regulations, and policy frameworks (Table 3.4).

3.1.7 Farming Systems in Uasin Gishu

According to Uasin Gishu County (2018), the County has a mixed farming system. The main cash crops are wheat and maize. Of the 132 ha of land under crops, 102, 753 ha is used for growing maize and 16, 867 ha for wheat. AIVs are grown on 132.8 ha. The main livestock reared in the County are beef and dairy cattle, sheep, goats, pigs, and chicken (exotic and indigenous). Uasin Gishu County (2018) further states that 200 million kgs of milk and 2.5 million kgs of meat are produced annually. Fish production has also been improved due to increased demand. A fish hatchery has been established in collaboration with UoE for constant supply of fingerlings to fish farmers in the County.

3.1.7.1 Quantitative Analysis

Uasin Gishu County (2018) observes that 132.8 h (332 acres) of land is under AIV production with approximately 112,000 smallholders producing AIVs. 80% of the smallholders are in rural areas, 15% in rural areas and 5% are contract farmers. Naivas Supermarket Chains have their own farms that produce vegetables, and they only purchase from other sources like private companies when they have deficits in supply and increased demand.

In the rainy season, 6000 kg of AIVs can be produced from an acre of land, per season of 3 months. Average farmgate prices are Ksh. 30/kg. Wholesalers and private companies purchase a kilo of AIVs from farmers at Ksh. 40 and Ksh 50, respectively. Private companies sell to small-scale processors, supermarkets, and high-end grocers at Ks. 80 and Ksh. 60, respectively. Small-scale processors need 2.5Kg–3 kg of fresh AIVs to produce 1 kg of processed AIVs that then retails for Ksh.400/kg. 100% of AIVs processed by large-scale processors is exported to Europe and the USA.

3.1.7.2 Calculating Value Shares

KIT et al. (2006) note that value share is the percentage of the final retail price that each actor manages to capture. Value shares were calculated at three levels in the AIV value chain in Uasin Gishu, as they are representative of the crucial levels and actors in the value chain. Calculations were done for yields per acre of land per season (3 months).

3.1.7.3 Production Level

Yield – 6000 kg Farmgate price = Ksh. 30/kg
Revenue = 6000 * 30 = Ksh. 180,000
Approximate Variable costs = Ksh. 60,000
Therefore, gross margin = 180,000–60,000 = Ksh. 120,000

3.1.7.4 Trading Level

Private companies and wholesalers
Average selling price – Ksh. 70 (80 + 60)/2
Revenue = 6,000 *70 = Ksh. 420,000
Approximate variable costs = Ksh. 150,000
Therefore, gross margin = 420,000–150,000 = Ksh. 270,000

3.1.7.5 Retail (High-End Supermarkets)

Selling price = Ksh. 90/kg
Revenue = 6,000 * 90 = Ksh. 540,000
Approximate variable costs = Ksh. 300,000
Therefore, gross margin = 540,000–300,000 = Ksh. 240,000 (Table 3.5)

Table 3.5 Value shares in AIV value chain in Uasin Gishu

Actor level	Variable costs (VC)	Revenue *(Rev) (selling price)*	Gross *income (Rev – VC)*	Value added *(Rev – previous actors Rev)*	Gross margin % *Gross income*100/ revenue*	Value share *added Value*100/ retail price*
Producers	60,000	180,000	120,000	180,000	66.67%	33.33%
Trading	150,000	420,000	270,000	240,000	64.28%	44.44%
Retailing (high end)	300,000	540,000	240,000	120,000	44.44%	22.22%

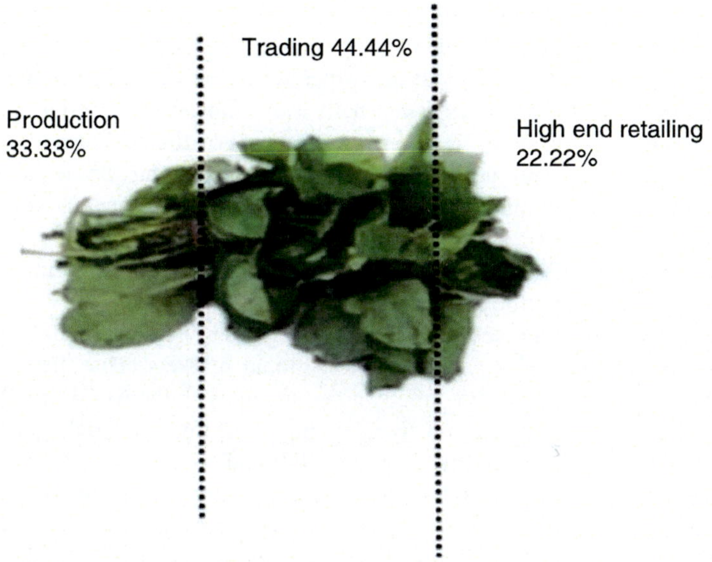

Fig. 3.5 Value share representation in AIV VC in Uasin Gishu

This can be represented as follows (Fig. 3.5).

3.1.7.6 Qualitative Analysis/Value Chain Governance

Market Links (2021) defines value chain governance as relationships among the buyers, sellers, service providers, and regulatory institutions that operate within or influence the range of activities required to bring a product or service from inception to its end use. Governance is about power and the ability to exert control along the chain – at any point in the chain, some firm (or organization or institution) sets and/or enforces parameters under which others in the chain operate. Table 3.1 highlights the present framework of the Uasin Gishu County AIV value chain. There is little or no coordination among the various actors in the VC, e.g., smallholder producers of AIVs must find their own markets, and training by KIRDI and KCIC is available to only a few smallholders selected through competitive processes. The chain actor relations are only coordinated when smallholders are dealing with buyers from the private sector, as this arrangement comes with drawing up of contractual obligations.

Presently there is no collective action from smallholder producers that is visible and/or yielding tangible results like negotiating for better prices and accessing financial services. Producer organizations formed (regional farmers associations or community-based groups) are usually temporary and formed for a specific purpose, e.g., accessing a onetime grant from donors. Even though the AIVs value chain has great market potential, it remains highly fragmented.

Supermarkets and high-end green grocery stores are the lead firms in the AIV value chain in Uasin Gishu. Due to the consumer demand for higher-quality AIVs, albeit at higher prices, these lead firms dictate the terms of the produce supplied. Private sector firms and wholesalers usually must transport the produce to the supermarkets and high-end stores. Payment terms are set by the lead firms and vary from weekly payments, monthly payments, or quarterly payments. Some payment terms may therefore pause liquidity problems for the suppliers, especially when the lead firm is also the largest buyer. The smallholder producers of AIVs are the most vulnerable actors, as they do not have the power to dictate the economic parameters concerning the production and marketing of AIVs.

There is one large processor in Uasin Gishu County, Mace Foods (Mace Foods, 2021). Mace Foods dries AIVs for international markets. They have contractual agreements with farmers who directly supply them with the AIVs. Small-scale processing in cottage industries involves sorting, washing, and packaging the AIVs. 99% of the processed vegetables is sold to high-end retail consumers and supermarkets due to the price differential that makes it out of reach for low-income consumers.

Deißler et al. (2021), in a study on gender dynamics in Kenya, note that in more than 70% of households producing AIVs, females are responsible for the production and sale of AIVs. They also note that the decisions regarding the income from AIV sales are still dominated by females, but the female bargaining power regarding the general crop production is still low. This suggests the need to empower the women involved in the AIV value chain. To promote a robust, resilient, and robust AIV value chain in Uasin Gishu, the following sustainability profile is proposed (Table 3.6).

Table 3.6 Sustainability profile of AIV value chain in Uasin Gishu County

People	Adequate production of AIVs by smallholders
Profits/ prosperity	Fair prices are given to producers of AIVs
Planet	Adopt environmentally sustainable production practices
Partnerships	Develop multi-stakeholder partnerships among the various stakeholders in the AIV value chain in Uasin Gishu County
Peace	Ensure the AIV value chain developed is dynamic and inclusive. This is especially important in Uasin Gishu which is a cosmopolitan county, prone to ethnic strife during political election cycles

3.1.7.7 Information Flows in the AIV Value Chain

The following information flows are present in the AIV value chain in Uasin Gishu County.

Product information – Demand is consumer driven and as such producers will produce AIVs as demanded by the consumers, via the lead firms.

Price information – Price flows and is managed at various levels of the VC. For example, middlemen will determine the prices they give to smallholder rural producers, and processors will determine what prices to charge for their finished products.

Market information – Uasin Gishu County sets the rates, fees, and levies to be paid by the different stakeholders in the value chain. The County Government also coordinates any external collaboration and funding opportunities available to farmers in the county. This information is usually communicated in newspapers and social media sites.

KIRDI and KCIC provide technical and value addition information to selected smallholders through the KIRDI Training Center and the AgriBiz Business Incubation Hub, respectively.

Even though the producers of the AIVs are price takers, there is information flow to other actors in the value chain on the products available and in what quantities, so that provision can be made in the event of shortage of surplus.

The information flows are therefore different at various levels in the value chain depending on the information provided. What is of note though is the high degree of information asymmetry in the value chain. The smallholder producers will usually be takers of market information as they do not have collective bargaining power.

3.1.8 Quality Management

Luning and Marcelis (2020) describe quality as meeting or exceeding customer and consumer expectations. Adapting from this definition and the consumer expectations in Uasin Gishu County, some quality attributes of AIVs include the following:

Fresh AIVs with leaves still green in color and not withered, and not having an unpleasant smell.

Cost – high-end consumers are willing to pay higher to get the exact specifications they need, while low-end consumers may not have the luxury to pay more for AIVs that are, for example, same-day harvested produce that is costlier.

Availability – AIVs are in more supply during the rainy season. This means the prices are lower during the rainy season and higher during the dry season.

Safety – high-end consumers are more concerned with the sources of their AIVs and will pay more for organic and AIVs that can be traced to source.

Perishability – AIVs have a short shelf life, and for actors along the value chain with no access to refrigeration facilities, this leads to postharvest losses.

Health benefits – AIVs have been touted to have a high nutritional value and this has increased their demand in the market.

Some factors that affect the quality of AIVs include pests and diseases, poor storage facilities leading to deterioration of the leaves, transporting them over long distances therefore exposing them to dirt and unhygienic conditions, no standardization standards for AIVs, and improper selling practices, e.g., selling by the roadsides, thus exposing the AIVs to bacteria and dust.

Quality control of food in Kenya is done by the Kenya Bureau of Standards (KEBS). Quality control is usually done at processing level. This therefore means only large-scale producers can have the mark of quality on products. Large-scale processors also provide seeds to their contracted farmers, and this ensures the production of quality produce. The process is also lengthy and expensive, meaning cottage processers of AIVs do not comply. This presents a gap in quality assurance in the AIV value chain in Uasin Gishu that needs to be addressed to ensure quality produce in the value chain.

3.1.9 Value Chain Analysis

3.1.9.1 Value Chain Map

See Fig. 3.6.

3.1.10 SWOT Analysis

A SWOT analysis highlights the strengths, weaknesses, opportunities, and threats in the AIV value chain in Uasin Gishu County. SWOT is a strategic planning tool as it aids in analyzing the internal and external factors that have an impact on a situation (Fig. 3.7).

3.1.11 PESTEC Analysis

The PESTEC analysis analyzes the external environment of the AIV value chain in Uasin Gishu County. There are several economic and technological factors the County can harness to develop the AIV value chain, e.g., opportunities to partner with other Counties and business incubation programs by KCIC (Figs. 3.8 and 3.9).

The causal diagram identifies the various issues in the AIV value chain in Uasin Gishu County. These include high transaction costs, distrust among stakeholders, no producer associations, and non-collaboration among the different actors in the value

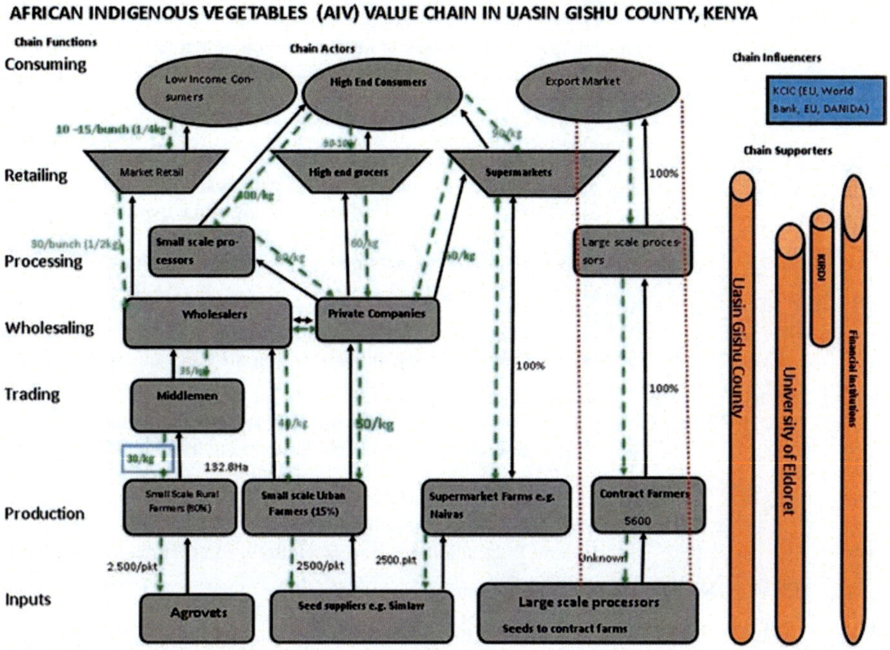

Fig. 3.6 Value chain map of AIVs in Uasin Gishu

STRENGTHS Surplus production during rainy season Increasing demand for AIVs Goodwill with the farmers and private sector	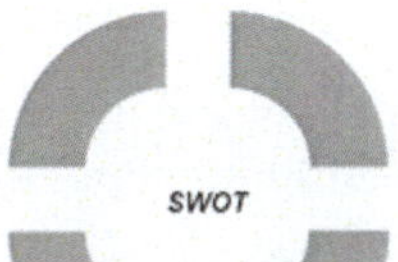 	WEAKNESSES Little coordination among the stakeholders in the AIVs VC High post-harvest losses due to perishability Distrust among producers and other stakeholders in the VC
OPPORTUNITIES Favorable climate for growing AIVs Access to Eldoret International Airport Increasing demand can increase prices to farmers if they are coordinated Value addition opportunities for AIVs – drying, fermenting, freezing Agribiz business incubation services for women and youth Good road infrastructure in urban areas in Uasin Gishu		THREATS Political instability in Uasin Gishu during election cycles Young people not interested in agriculture because it is deemed to be unglamorous High input costs Inefficient extension services Information asymmetry – smallholders lack control to market information.

Fig. 3.7 SWOT analysis of AIV value chain in Uasin Gishu County

P	POLITICAL	Uasin Gishu County coordinates agriculture policy Fees, levies and taxes to farmers and other actors in the VC Insufficient County support to smallholders Political strife in Uasin Gishu during election cycles
E	ECONOMIC	High cost of inputs Insufficient credit from banks Rising number of middle class in Uasin Gishu Supermarkets and high-end grocers control the market Chance to participate in contract farming for large processors NOREB – offers partnership opportunities with other counties in the NorthRift region
S	SOCIAL	Low involvement of youth in agriculture Women have low bargaining power in the AIVs chain
T	TECHNOLOGICAL	Agribiz business incubation hubs for select smallholder Value addition e.g., fermenting, and drying training offered by KIRDI Availability of hybrid seeds Irrigation by urban smallholders Eldoret International airport is a hub for international markets.
E	ENVIRONMENTAL	Suitable climate to grow AIVs. Ability to grow AIVs 3 times in a year Several types of AIVs can be grown together and inter-cropped with other crops.
C	CULTURAL	Uasin Gishu is a cosmopolitan county presenting diverse market for AIVs AIVs are being embraced for their nutritious properties Traditional Indigenous knowledge used e.g., in fermentation of AIVs.

Fig. 3.8 PESTEC analysis of AIV value chain in Uasin Gishu

chain. This leads to the ultimate problem in the chain – uncoordination among the actors in the AIV VC in Uasin Gishu County.

3.2 Conclusion

The purpose of this study was to analyze the African Indigenous Vegetable (AIV) value chain (VC) in Uasin Gishu County to identify the problem and indicate two directions for chain development.

The study identified the main problem in the AIV VC as uncoordination among the various actors in the value chain. There is little or no collaboration among the various actors in the value chain, no producer organizations, and high transaction costs. A coordinated value chain is predicated on trust among the various actors in

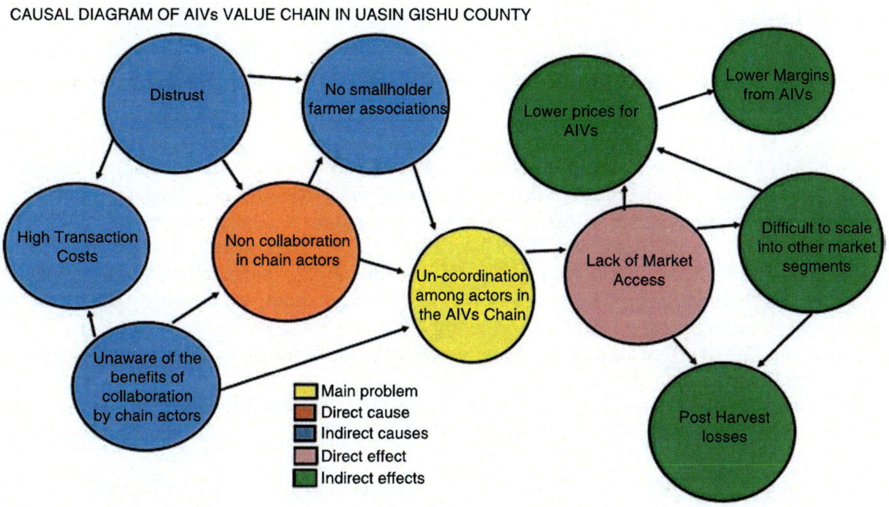

Fig. 3.9 Causal diagram of AIV VC in Uasin Gishu County

the value chain. A lack of trust leads to individualistic behaviors like farmers having to find market for their own produce and a lack of economies of scale along the value chain. This leads to ineffective and inefficient market access by the actors leading to lower prices, postharvest losses, inability to scale market activities, and low margins for different players along the value chain.

3.2.1 Recommendations

To ensure coordination on the African Indigenous Vegetable Value Chain in Uasin Gishu County, efforts would need to be expended to ensure there are coordinated efforts along the various actors in the VC. Two specific recommendations that would allow for this to be realized include:

Benchmarking and gap assessment with Vihiga County that is already implementing a commercialization of AIVs in their operations. This will allow Uasin Gishu to identify the gap areas, and how best to address them in context.

Ensuring market access for smallholders by ensuring they are fully included in the VC processes and dynamics. This can be achieved by first ensuring the smallholders are aware of the benefits of collaboration and being organized. This can be achieved through the already existing private sector collaborative efforts.

Ultimately, these will ensure smallholder producers are included in the AIV VC and are able to fully collaborate with other actors to ensure a VC that is coordinated.

References

Abukutsa, M. (2021). Experts call for more synergy to support optimum use of AIVs Available at: https://www.jkuat.ac.ke/experts-call-for-more-synergy-to-support-optimum-utilization-of-aivs/. Accessed 30 Nov 2021.

Alulu, J., Otieno, D., & Kosura, W. (2019). Drivers of transformations in smallholder indigenous vegetable value chains and contract farming evolution in Western Kenya, paper presented at 6th international conference of African agricultural economists, Abuja, September 23–26, 2019. Available at: https://ideas.repec.org/p/ags/aaae19/295739.html. Accessed 1 Dec 2021.

Council of Governors. (2021). Vihiga County commercializing indigenous vegetables. Available at: https://www.cog.go.ke/47-county-news/530-vihiga-county-commercialising-indigenous-vegetables. Accessed 2 Dec 2021.

County Guide. (2021). Uasin Gishu County. Available at: https://www.kenyacountyguide.co.ke/uasin-gishu-county/. Accessed 24 Nov 2021.

Deißler, L., Krause, H., & Grote, U. (2021). Gender dynamics and food security in the Kenyan African Indigenous Vegetables Supply Chain. Available at: https://ideas.repec.org/p/ags/iaae21/314983.html. Accessed 2 Dec 2021.

Info Track. (2021). Uasin Gishu County. Available at: http://countytrak.infotrakresearch.com/uasin-gishu-county/. Accessed 23 Nov 2021.

Kebede, S. W., & Bokelmann, W. (2017) African indigenous vegetables and their production practices: Evidence from the HORTINLEA Survey in Kenya. *Agrotechnology, 6*, 170. https://doi.org/10.4172/2168-9881.1000170

KIT, Faida MaLi, & IIRR. (2006). Chain empowerment: Supporting African farmers to develop markets. Royal Tropical Institute, Amsterdam; Faida Market Link, Arusha; and International Institute of Rural Reconstruction, Nairobi. kit.nl/smartsite.shtml?id=6768. Accessed 2 Dec 2021.

Luning, P., & Marcelis, W. (2020). *Food quality management: Technological and managerial principles and practices* (pp. 24–25). Wageningen Academic Publishers. Accessed 2 Dec 2021.

Mace Foods. (2021). *About us*. Available at http://www.macefoods.com/welcome/aboutus. Accessed 2 Dec 2021.

Market Links. (2021). Value chain governance overview. Available at: https://www.marketlinks.org/good-practice-center/value-chain-wiki/value-chain-governance-overview. Accessed 2 Dec 2021.

Mili, S. (2017). Benchmarking agri-food value chain performance factors in South Mediterranean Countries. Available at: https://www.semanticscholar.org/paper/Benchmarking-Agri-Food-Value-Chain-Performance-in-Mili/87b5dcf169202955ae0bfc21ee83a63d1138d335. Accessed 2 Dec 2021.

Modi. (2021). Indigenous foods knowledge in the curriculum. Available at: https://ndabaonline.ukzn.ac.za/UkzndabaStory/Vol8-Issue29/DVC%20Weighs%20in%20on%20the%20Place%20of%20Indigenous%20Foods%20in%20the%20Curriculum/. Accessed 1 Dec 202.

Ngenoh, E., Kurgat, B., Bett, H., Kebede, S., & Bokelmann, W. (2019). Determinants of the competitiveness of smallholder African indigenous vegetable farmers in high-value agro-food chains in Kenya: A multivariate probit regression analysis in *Agricultural and Food Economics* 7:2. Available at: https://agrifoodecon.springeropen.com/articles/10.1186/s40100-019-0122-z. Accessed 2 Dec 2021.

Otieno, A., Obel, C., Ongwen, S., Omanga, S., & Bokelmann, W. (2019). The African indigenous vegetables value chain governance in Kenya in *Studies in Agricultural Economics 121* (2019) 41–52. Available at: https://doi.org/10.7896/j.1818. Accessed 2 Dec 2021.

Rampa, F., Lammers, E., Linnemann, A., Schoustra, S., & Winter, D. (2020). The promise of African indigenous foods and technologies Pathways to improved food and nutrition security of the poor. Available at: https://research.wur.nl/en/publications/pathways-to-improved-food-and-nutrition-security-of-the-poor-the-. Accessed 2 Dec 2021.

Uasin Gishu County. (2018). Country integrated development plan (CIPD) 2018–2022. County Treasury, Eldoret. Accessed 15 Nov 2021.

World Bank. (2021). Public Finance Management Act 2012. Available at: https://www.world-bank.org/content/dam/Worldbank/document/Africa/Kenya/Kenay%20Devolution/Public%20Finance%20Management%20Act%20%282012%29.pdf. Accessed 2 Dec 2021.

Chapter 4
Assessing the Physical Contamination of Food through the Use of Metallic Staple Pins during Packaging with the Kenyan Market as a Case Study

Faith Onyangore

4.1 Introduction

The food market has been proliferating over the existence of human civilization with the demand for different products equally growing. The methods of distribution have also evolved along with the changing demands and requirements by consumers. For instance, the general market has grown from the basics of selling uncooked farm products into processed and fast foods that are dominating the consumption spectrum. As a result of such changes, food packaging has become an important part of the market as a way of ensuring adequate distributions in carriage and hygiene. Various concerns surround the food packaging technologies that involve contamination of the foodstuffs and sometimes extend to the pollution of the environment. For example, plastic packaging has wreaked havoc onto most ecosystems since it is mainly nonbiodegradable (Mwaura et al., 2016). The concerns for this study involve the use of packaging materials that are likely to contaminate the foodstuffs. This issue emerges in light of the growing use of metallic staple pins for food packaging mainly in the Kenyan market. This aspect has gained popularity after the plastic ban that saw the vendors shift toward paper-based packs that require staples. The goal is to establish the contamination effects, the forces that motivate the use of this method, and the general view among vendors and consumers regarding the use of staple pins for food packaging with a case focus on the Kenyan market.

Food safety is an issue of significant concern for any government or community. This aspect is specifically important when considering the distribution of food through primary manufacturers and secondary producers. The Kenyan food market is particularly dynamic through its diversity of producers ranging from domestic cooks to restaurant and street vendors and even industrial manufacturers (Hoffmann

F. Onyangore (✉)
University of Kabianga, Eldoret, Kenya

S. D. Carter (ed.), *COVID-19, Supply Chain, Climate Change, and Sustainable Development in Africa*, Springer Proceedings in Business and Economics,
https://doi.org/10.1007/978-3-031-26121-3_4

& Moser, 2017). Various practices exist to satisfy this diversity in terms of the demands of the consumers of the relevant products. However, food safety is of paramount significance regardless of the producers, distributors, or consumers of the foodstuffs (Grace, 2015). This aspect is important since food safety practices protect the consumers from different physical and biological harms that emerge from contaminants (Corburn & Hildebrand, 2015). Different forms of contamination threaten the aspect of food safety among those participants. Physical contaminants are some of the most common foreign materials that pose threats to the consumers (Malavi et al., 2018). Such contaminants can enter the foods during preparation or even packaging. This study particularly focuses on the issue metallic staple pins as a contaminant during food packaging in the Kenyan food market.

Previous research has highlighted the various issues relating to food contamination, but has not focused intensively on the prevalence of the problem in the Kenyan market. Equally, insufficient focus has gone toward ensuring public awareness about this problem despite the threat that it presents to the people. For instance, most studies have emphasized the use of staple pins in packaging at an industrial level (NDTV, 2019). However, very few studies have highlighted the problem outside the industrial realm into other food markets such as roadside vendors and small-scale restaurants. This study aims at providing an extensive analysis of the various realms, industrial and nonindustrial, where the use of staple pins for packaging is likely to present a threat. This investigation intends to provide sufficient information that can enhance the creation of policies and other solutions that can curb the practice in its entirety of reach by highlighting the threats that staple pins present in their areas of food packaging applications.

4.1.1 Understanding the Physical Contamination of Food

Hygiene is an essential component of food consumption as a means for ensuring the safety of the consumers. Various activities are likely threats to the foods that people consumes, which necessitates the observation of strict routines to prevent such hazards. However, some other actions can still pose a threat even though the actors did not intend it. The physical contamination of food involves the introduction of foreign matter into a foodstuff in a manner that may cause either intentional or unintentional harm to the consumer (Ndaramu et al., 2019). For example, the presence of human hair in a plate of food is contamination while adding tomato sauce to a plate of potato chips is not. Both scenarios involve the introduction of foreign matter into a foodstuff, but the former is harmful, thereby becoming a form of physical contamination. The physical contaminants can also be nonsolid in certain scenarios (Bereda et al., 2016). For instance, if dirty water leaks into a cooking pot of food, then it constitutes a form of physical contamination (Tan et al., 2016). Such actions of contamination indicate an improper emphasis of hygiene as a cause for concern (Ismail et al., 2016). However, other hygienic actions such as inappropriate food packaging can pose a threat to the consumers (Malavi et al., 2018). The use of staple pins along with paper-based packages in foodstuffs is an example of such contaminants (Musyoka et al., 2018). Understanding such issues is essential to offer guidelines to both vendors and consumers of the risk factors and their hazards.

Physical contamination of food can be a result of different activities relating to the production processes. For instance, improper storage of food products or the raw materials that facilitate their product can be a common cause for the introduction of foreign matter (Ndaramu et al., 2019). An example is when rats or other pests can infiltrate the storage facility and bring in different contaminants into the foodstuff (Kariuki et al., 2017). Other cases of contamination can relate to the processes of preparation of handling of the food products both in the markets and even domestically (Ndaramu et al., 2019). Such aspects relate to the hygienic routines that the handlers utilize in their production and how it aims at eliminating the risks of contamination. For example, improper cleaning of preparation facilities or the inadequate removal of contaminants from a workplace is a cause for concern (Ismail et al., 2016). Physical contamination can also relate to improper packaging equipment or materials that may contaminate the food contents (Garvey, 2019). For example, the use of unclean containers for packaging or holders made of harmful materials can be risky to the consumers through different forms of contamination. For instance, some metallic containers can include heavy metals such as zinc that are harmful to humans (Massadeh et al., 2018). Stringent assessment is then necessary in the food processing chain to prevent physical contamination and safeguard the consumers.

4.1.2 The Common Uses of Staple Pins as a Packaging Product

The packaging of food in different forms has become an essential part of the modern life to meet consumer demands. This practice enables people to access various types of foodstuffs conveniently with hygiene as another element necessitating packaging (Tshipamba et al., 2018). As a result, the use of plastic and paper bags requires different methods of sealing to ensure the safety of the contents (Hammad, 2016). Ideally, thermal sealing of safe adhesives is the most suitable approach in food packaging as they adhere to different safety regulations and reduce risks to consumers (Marichelvam et al., 2019). However, innovative vendors and food producers are equally eager to pursue other alternatives to satisfy the demand of their consumers (Bellia et al., 2016). For instance, metallic staple pins are a common feature in various forms of food packaging along with plastic and paper bags (Musyoka et al., 2018). This practice has proliferated in most areas as a convenient sealing method that is cost-effective and equally quick to meet the requirements of both the sellers and their buyers (Bagrecha, 2018). The use of staple pins is also pertinent to various manufacturers of food products. A review of the common areas of application for metallic staple pins in the food industry can highlight their prevalence and potential risks as a part of this investigation.

Fast foods are gaining popularity in the contemporary society as a convenient choice to fit into most busy lifestyles. A common sight involves takeaway foods from different joints as a means of keeping up with the contemporary fast life. In Kenya, for example, foodstuffs like potato chips, fried chicken, crisps, and other deep-fried items are common purchases from fast-food joints (Njite et al., 2015). The sales of such items requires fast, cost-effective, and convenient packaging to meet the requirements of both the buyers and sellers (Musyoka et al., 2018). As a result, paper and plastic bags are the most ideal packaging choice for most vendors as they meet all those aspects (Elisha et al., 2016). However, the fastest way for sealing the packaged foods differs depending on the specific materials that the vendors use. For instance, most vendors previously used plastic bags that they tied a knot at the top as a sealant before the ban on this method of packaging (Kariuki et al., 2017). The adoption of paper-based bags for food packaging has motivated the use of staple pins as the most convenient sealant choice for most vendors (Horvath et al., 2018). Alternatives like sealant tapes or glue have not caught up in most markets since they take time to seal the products, which contradicts the needs of the fast-food industry. As a result, fast-food joints have become one of the most prominent users of metallic staple pins as a packaging sealant in their processing.

Teabags are a convenient tool in most modern households since they eliminate the need for sieves and other utensils for the users. Manufacturers of teabags use different approaches for sealing the products for efficient use among the consumers (Kipkirui et al., 2017). Staple pins as a sealing method for teabags is a common sight in most global markets despite the risks that they pose (NDTV, 2019). The manufacturers use those staple pins to hold the immersion strings in place as a convenient way for consumers to hold the teabags while dissolving them into the tea (NDTV, 2019). Another use of the staples in these products includes on the packaging of the teabags to keep the items in their covers before delivery to consumers (Bagrecha, 2018).While uncommon in most areas, this practice is prominent in other regions where regulations have failed to curb it, which threatens the consumers that have to face it. These aspects indicate the use of staple pins as a method for packaging and utilizing teabags despite the potential risks that they pose to the users.

4.1.3 Factors Motivating the Use of Staples in Food Packaging

The etymology of the use of staple pins in food packaging derives from various factors that motivate both vendors and producers. For instance, cost is a primary driver for the application of staple pins as a method for sealing food packages in most applications (Malavi et al., 2018). This aspect coincides with the ease of access to the staples in the market as they are readily available as a part of stationery (Malavi et al., 2018). The cost-effective aspect is suitable for both primary manufacturers and secondary producers, such as fast-food joints, where the target packaging is mostly in single-use basis. For example, the purchase of potato chips in Kenya involves packaging that the buyer will use only once and dump afterward (Abong &

Kabira, 2015). This aspect motivates producers to use cost-effective approaches that will enable them to cut down on production expenses and maintain high returns (Brennan et al., 2017). Equally, metallic staple pins are easy and quick to stick to the packaging, which motivates its use even in stationery applications (Attokaran, 2017). For example, it is easier and quicker to use a staple pin on a paper-based package than it is to apply glue since the latter takes longer to adhere and secure the contents (Abong & Kabira, 2015). Therefore, the aspects of cost and ease of use motivate most vendors and manufacturers to use staple pins in their products rather than applying safer alternatives.

Environmental and legal factors also have a significant role in motivating the use of staple pins in food packaging. Primarily, most authorities have paid insignificant attention to the risks that relate to such practices in the food industry, which has led to the absence of laws to regulate the relevant practices (Hoffmann & Moser, 2017). A public that is equally unaware of such issues has an inconsiderable concern over the risks, which implies that inadequate action goes toward seeking safe alternatives. In Kenya, the ban on plastic bags has played a key role in the proliferating use of staple pins in the packaging of fast foods (Horvath et al., 2018). The plastic ban led to the increase in the use of paper-based bags as an environmentally friendly alternative to adhere to the new laws (Mwaura et al., 2016). However, the new packaging method also required the vendors and other users to develop appropriate strategies to cope with market demands (Horvath et al., 2018). As thermal sealing is inapplicable on paper-based products, staple pins emerged as the most suitable approach for quick and easy packaging (Uchida et al., 2018). These aspects indicate that the sociopolitical and environmental factors in the market are equally influential in motivating the use of staple pins as a method for food packaging.

4.1.4 Hazards Emerging from the Use of Staples in Food Packaging

The presence of staple pins in food packaging poses various dangers to the consumers of those products. The primary epidemiology of risks involves the probable cause of injury to users in various forms. For example, the small size of staples is a cause for concern since most consumers may fail to notice those pins when they drop from the packaging into the foodstuffs (Coles, 2015). As a result, the consumers are likely to chew on the pins while they are eating those foods, which is a direct cause for injury to the mouth. This risk is particularly prominent among small children who might not pay attention to the potential dangers of the staple pins in their food (TNN, 2017). Still, adults equally need to watch out for the accidental presence of the metallic items in their foodstuffs. The risk equally involves the probable ingestion of the metallic pins along with the food for the consumers (Coles, 2015). The risk of ingesting staple pins is particularly pertinent to teabags where the consumer may swallow the drink without realizing its presence (James et al., 2018).

This risk escalates from the fact that the consumer will not chew the drink like they would other foods, which increases the danger of ingestion (TNN, 2017). Therefore, the presence of metallic staple pins in food packaging poses injurious risks to consumers in terms of chewing and ingestion in the probable instance that they drop into the foodstuffs.

Various complications are likely to arise from the injurious harms that emerge from the ingestion of staple pins. For example, this risk is particularly prominent among small children who might not pay attention to the potential dangers of the staple pins in their food (TNN, 2017). Still, adults equally need to watch out for it involves a delicate part of the human body where the victim is unable to control the movement of the food and the contaminant (Abolfotouh & Moore, 2019). As a result, it is a significant medical concern that should discourage the use of staple pins in packing foodstuffs due to the risks that it creates for the consumers (Modikoe et al., 2019). The pins can equally be injurious to the person that ingests them while they are in the stomach since they maintain their solid state that is harmful due to their sharp pointed ends (Law et al., 2019). Therefore, the victim should seek medical attention to prevent the effects of potential internal bleeding. These aspects indicate the probable physical injuries that are likely to arise in the event that a person swallows a staple pin, and the presence of those items in food packaging increases the risk.

Different nonphysical harms are equally probable as a result of the ingestion of staple pins for food consumers. The primary concerns emerge from the items' chemical composition relating to the galvanization process as a means of preventing corrosion on the pins (Luo et al., 2019). This risk is particularly harmful to consumers of teabags whose consumption involves introducing the pins directly into their drinks and heating them (TNN, 2017). The presence of the zinc that manufacturers use to galvanize the pins in a high-temperature environment increases the potential for the metal to dissolve into the drink. Such heavy metals are a common cause for digestive complications and other medical problems that may even lead to death (Sankhla et al., 2016). The galvanization process also implies that the consumer is unable to digest the pins, and absorbs the heavy metals into their bloodstream instead. This action increases the risks of physical injury besides the obvious chemical contamination to the consumers (Abolfotouh & Moore, 2019). While ingestion in smaller quantities may not pose any serious risks on the chemical contamination perspective, continuous exposure to this threat is a matter of significant concern. Assessing the most appropriate remedies to the problem is essential to protect the consumers from the harms relating to the utilization of metallic staple pins in food packaging.

4.1.5 Remedies and Alternatives to the Use of Staple Pins for Food Packaging

Various approaches are necessary to ensure the safety of food consumers amidst concerns about the risks about the use of staple pins in food packaging. Primarily, public sensitization is the most appropriate action to undertake in such scenarios (Mbae et al., 2018). Efforts are necessary to inform citizens and other consumers about the risks that can emerge from the presence of those metallic pins in their food packaging. This undertaking is specifically useful in areas where regulations are not vigilant enough to tackle and curb the practice of using staple pins as a method for sealing food packaging (Tsakali et al., 2016). For instance, the Kenyan government has undertaken minimal effort toward this issue since it has not arisen in any consumer safety discussions. As a result, the country does not have any policies to regulate the practice despite its widespread use in the local market. This aspect indicates the need to inform the public of the dangers that relate to those staple pins in their food packaging and the need for their vigilance in preventing any relevant harms. For example, sensitization can enable a consumer to deny the utilization of this method when they are purchasing from the vendor. The goal for public sensitization is to bridge the policy gap and steer consumer preferences in a direction that will cause the use of staple pins in food packaging among sellers and manufacturers.

Most governments have established agencies that assist them in regulating different production actions and methodologies in the food industry. However, the efficiency of the food regulation agencies relies intensively on the presence of stringent policies as primary guidelines for such undertakings (NDTV, 2019). As a result, the Kenyan government needs to improve its efforts toward the creation of regulations to curb the practice of using staple pins in food packaging. Similar efforts in the plastic ban have seen a decline in the use of that material in the Kenyan market as an indicator of the role that policies play in ensuring public safety (Mwaura et al., 2016). Other countries have seen the successful implementation of policies prohibiting the use of staple pins in food packaging in both primary and secondary producers in their economies (NDTV, 2019). For instance, India introduced a ban on the use of metallic staples in food packaging in the Food Safety and Standards (FSS) Act of 2006 (TNN, 2017). This move was pertinent to the prominent use of staple pins in fast foods and teabags that posed a significant threat to users, especially children (NDTV, 2019). Despite some pushback from producers, the success of the ban in the FSS Act is an indication that adequate regulations can be a sufficient remedy to the threats that staple pins pose in food packaging applications.

Another prominent approach to remedying the use of staple pins for food packaging involves incentivizing and regulating consumers to develop alternatives to this method. This aspect requires the localization of the solution to the specific industry requirement and the context of application. For example, teabag manufactures can use glue sealant and thread knots as alternatives to staple pins in their respective applications (TNN, 2017). This move is specifically useful to eliminate risks that involve the ingestion of the metallic items that can cause medical harm to consumers. Fast-food joints and other similar secondary producers like malls and grocers should consider alternatives like thermal sealing to reduce their dependency on metallic staple pins for food packaging (Marichelvam et al., 2019). This approach is specifically useful for sealing products where the packaging is plastic bags. Quick drying glue is another alternative that producers using paper-based packaging should consider in their processes as a safer alternative to metallic staple pins (Sanyang et al., 2016). Both approaches aim at eliminating the need for physical risks to the foodstuffs and reduce the dangers relating to the ingestion of the sealant or packaging materials. Therefore, assessing and localizing various alternatives is the best methodology to resolve the risks relating to the use of staple pins in the packaging of food products.

4.2 Conclusion

An evaluation of the use of staple pins in the Kenyan market for food packaging reveals various aspects whose understanding is essential to establish the risks and remedies. While the approach is a cost-effective method for both vendors and consumers of food products, it poses various threats, which necessitates and environmental assessment to determine the motivating factors and remedies. The evaluation determines that metallic staple pins are a potential physical contamination to the foodstuffs in which vendors use the items for packaging. The common application of this practice involves primary manufacturers like those making teabags and secondary producers such as fast-food joints, retailers, malls, and grocers. The risks that relate to the physical contamination of staple pins include injuries to the mouth during chewing and scratches that can cause bleeding to both the food pipe and intestinal systems. The galvanization on staple pins is equally harmful since it contains heavy metals whose ingestion is a serious chemical contaminant and risk to the human body. The remedies include creating public awareness about the problem as a way of enabling the citizens to overcome the challenge. This move can also motivate producers to seek alternatives to staple pins by localizing their solutions to the relevant industrial needs. Policies are also an essential approach to regulate the market and encourage the use of safe food packaging methods. The overall observation is that staple pins pose a significant threat to consumers in their application as a food packaging method and producers should instead pursue appropriate alternatives.

4.2.1 Future Prospect

This study highlights the challenges that emerge from the utilization of metallic staple pins in food packaging as a risk in contaminating the consumables in the Kenyan market. The assessment highlights these problems in light of the application of those pins in manufacturing and other secondary food packaging actions such as through street vendors. This evaluation concludes that actions are necessary to assist the Kenyan populace in overcoming the problem. However, further investigations are necessary to establish the existing alternatives that those users can utilize. The most important part in this future research is to determine the safest alternatives that meet the requirements for various packaging actors in terms of cost and quickness to use. For instance, the prospective studies should analyze packaging methods that can enable fast-food joints and street vendors of food to cost-effectively package their goods while ensuring sufficient speed to satisfy the waiting customers. Those studies also need to assess those alternatives in terms of their environmental impact as well as contamination potential in a manner that might threaten human health. Those investigations are an equally useful step in alleviating the current use of staple pins by providing the most appropriate alternatives.

References

Abolfotouh, S., & Moore, D. (2019). Use of simultaneous traction over a halo ring to achieve reduction of a type 2 odontoid fracture for anterior odontoid screw fixation. *International Journal of Surgery Case Reports, 64*, 97.

Abong, G. O., & Kabira, J. N. (2015). Potential food safety concerns in fried potato products in Kenya. *Open Access Library Journal, 2*(05), 1.

Attokaran, M. (2017). *Natural food flavors and colorants*. Wiley.

Bagrecha, P. P. D. C. (2018). Business opportunities in small-scale Industries for Entrepreneurs in India. *BIMS International Research Journal of Management and Commerce, 3*(1).

Bellia, C., Pilato, M., & Seraphin, H. (2016). Street food and food safety: A driver for tourism? *CAL, 17*(S1), 20.

Bereda, T. W., Emerie, Y. M., Reta, M. A., & Asfaw, H. S. (2016). Microbiological safety of street vended foods in Jigjiga City, Eastern Ethiopia. *Ethiopian Journal of Health Sciences, 26*(2), 163–172.

Brennan, M., Frey, D., Goentzel, J., & Sundar, P. (2017). Reports from CITE "Food aid packaging evaluation in Djibouti, Ethiopia & South Africa".

Coles, C. (2015). A drawing pin, drill bit, several staples and a magnet: Definitely not a simple case of appendicitis. *BMJ Case Reports, 2015*, bcr2015211638.

Corburn, J., & Hildebrand, C. (2015). Slum sanitation and the social determinants of women's health in Nairobi, Kenya. *Journal of Environmental and Public Health, 2015*, 1.

Elisha, G. O., Arnold, O. M., Christian, U., & Huyskens-Keil, S. (2016). Postharvest treatments of African leafy vegetables for food security in Kenya: A review. *African Journal of Horticultural Science, 9*, 32.

Garvey, M. (2019). Food pollution: A comprehensive review of chemical and biological sources of food contamination and impact on human health. *Nutrire, 44*(1), 1.

Grace, D. (2015). Food safety in low and middle income countries. *International Journal of Environmental Research and Public Health, 12*(9), 10490–10507.

Hammad, J. F. (2016). U.S. Patent No. 9,352,864. U.S. Patent and Trademark Office.

Hoffmann, V., & Moser, C. (2017). You get what you pay for: The link between price and food safety in Kenya. *Agricultural Economics, 48*(4), 449–458.

Horvath, B., Mallinguh, E., & Fogarassy, C. (2018). Designing business solutions for plastic waste management to enhance circular transitions in Kenya. *Sustainability, 10*(5), 1664.

Ismail, F. H., Chik, C. T., Muhammad, R., & Yusoff, N. M. (2016). Food safety knowledge and personal hygiene practices amongst mobile food handlers in Shah Alam, Selangor. *Procedia-Social and Behavioral Sciences, 222*, 290–298.

James, V., Hamzah, H. B., & Ganapathy, S. (2018). Handheld metal detector screening for metallic foreign body ingestion in children. *Journal of Visualized Experiments, 139*, e58468.

Kariuki, E. N., Waithera Ng'ang'a, Z., & Wanzala, P. (2017). Food-handling practices and environmental factors associated with food contamination among street food vendors in Nairobi County, Kenya: A cross-sectional study. *EA Health Research Journal, 1*(1), 62–71.

Kipkirui, J. B., Ayienga, E., Milimu, M. B., Ndegwa, S. M., Ng'eno, V. K., Mogaka, M. D., Githiomi, J., Magiri, G. M., Ochieng, J. A., Osiako, J. O., & Kiilu, P. M. (2017). The effect of innovation strategies on market share of small scale tea packers in Kenya.

Law, K., Peck, J., Williams, M., Rippel, S., Wilsey, A. D., Swan, E. K., Falconer, T. G., & Wilsey, M. (2019). Penetrating antral straight pin injury following accidental ingestion: A case report. *Acta Scientific Gastrointestinal Disorders, 2*(3).

Luo, G., Li, H., Li, Y., & Mo, J. (2019). Microstructures and properties of a low-carbon-chromium ferritic stainless steel treated by a quenching and partitioning process. *Materials, 12*(10), 1704.

Malavi, D. N., Muzhingi, T., & Abong, G. O. (2018). Good manufacturing practices and microbial contamination sources in Orange fleshed sweet potato puree processing Plant in Kenya. *International Journal of Food Science, 2018*.

Marichelvam, M. K., Jawaid, M., & Asim, M. (2019). Corn and Rice starch-based bio-plastics as alternative packaging materials. *Fibers, 7*(4), 32.

Massadeh, A. M., Allah, A., & Al-Massaedh, T. (2018). Determination of heavy metals in canned fruits and vegetables sold in Jordan market. *Environmental Science and Pollution Research, 25*(2), 1914–1920.

Mbae, K. M., Ndwiga, M. K., & Kiruki, F. G. (2018, 2018). Microbiological quality of kachumbari, a raw vegetable salad popularly served alongside roast meat in Kenya. *Journal of Food Quality*.

Modikoe, M. M., Reid, M., & Nel, R. (2019). Oral health-related knowledge, attitudes and practices of adult patients in Mangaung metropolitan municipality, South Africa. *South African Dental Journal, 74*(4), 180–186.

Musyoka, J., Njunge, F., & Muzhingi, T. (2018). Food safety training workshop. Theme: Managing food safety and quality in small-scale food processing for roots, tubers and bananas (RTB) value chains in Sub-Saharan Africa.

Mwaura, A. W., Letting, N., Ithinji, G. K., & Bula, H. O. (2016). Green distribution practices and competitiveness of food manufacturing firms in Kenya.

Ndaramu, G. O., Makopondo, R. O., & Kariuki, A. (2019). Investigation on conformance of catering and food safety practices at the National Youth Service in Kenya. *Studies, 1*(3), 18–34.

NDTV. (2019). FSSAI bans use of staple pins in tea bags from June 30; Rules out Possibility of Extension. *NDTV Food*. Retrieved from https://food.ndtv.com/food-drinks/fssai-bans-use-of-staple-pins-in-tea-bags-from-june-30-rules-out-possibility-of-extension-1986244

Njite, D., Njoroge, J., Parsa, H., Parsa, R., & van der Rest, J. P. (2015). Consumer patronage and willingness-to-pay at different levels of restaurant attributes: A study from Kenya. *Research in Hospitality Management, 5*(2), 171–180.

Sankhla, M. S., Kumari, M., Nandan, M., Kumar, R., & Agrawal, P. (2016). Heavy metals contamination in water and their hazardous effect on human health—A review. *International Journal of Current Microbiology and Applied Sciences, 5*(10), 759–766.

Sanyang, M. L., Sapuan, S. M., Jawaid, M., Ishak, M. R., & Sahari, J. (2016). Effect of plasticizer type and concentration on physical properties of biodegradable films based on sugar palm (Arengapinnata) starch for food packaging. *Journal of Food Science and Technology, 53*(1), 326–336.

Tan, E. Y., Arifullah, M., & Soon, J. M. (2016). Identification of Escherichia coli strains from water vending machines of Kelantan, Malaysia using 16S rRNA gene sequence analysis. *Exposure and Health, 8*(2), 211–216.

TNN. (2017). Stop using staple pins to pack food: FDA. *Times of India*. Retrieved from https://timesofindia.indiatimes.com/city/goa/stop-using-staple-pins-to-pack-food-fda/articleshow/57583616.cms

Tsakali, E., Gortzi, O., Timpis, D., & Tsaknis, J. (2016). Food safety and quality control in the public catering sector-intervention programs. *Nutrition and Food Technology, 2*(3).

Tshipamba, M. E., Lubanza, N., Adetunji, M. C., & Mwanza, M. (2018). Evaluation of the effect of hygiene practices and attitudes on the microbial quality of street vended meats sold in Johannesburg, South-Africa. *Journal of Food: Microbiology, Safety & Hygiene, 3*(137), 2476–2059.

Uchida, K., Uematsu, I., Hayamizu, N., Tokuno, Y., Naka, T., Tanaka, M., & Terada, T. (2018). U.S. Patent Application No. 15/556,918.

Part III
Sustainable Development

Chapter 5
A Framework to Include Sustainability in Consumer Protection Policies

Laura Best and Miemie Struwig

5.1 Introduction

Democratic governments, once elected, set up and run an administration system which provides the state machinery to implement delivery of the political intentions as promised to the electorate. The government then develops policy frameworks to manage the socio-economic and political realities of the societies they have been elected into power to govern (Waller et al., 2014; Gumede, 2008). Consumer protection policy is one such domain of public policy. Recognizing "*that consumers often face imbalances in economic terms, educational levels, and bargaining power*" in relation to businesses and corporates when buying goods and services, the United Nations has issued Guidelines for Consumer Protection (UNGCP) (United Nations Guidelines, 2003). As a regulatory measure, the United Nations encourages governments to "*develop or maintain a strong consumer protection policy*" (United Nations Guidelines, 2003) while doing this within the context of countries' respective socio-economic and environmental conditions and taking cognizance of the needs and expectations of the people living in that country.

Consumer protection policies do not necessarily modify consumer behavior in terms of the extent of consumption. By not regulating this dimension, consumers are unrestricted in how much they consume. This is in contrast to another domain of public policy, namely, sustainability policy, which encourages reduced consumption owing to environmental concerns and the future impact on the earth's resources.

L. Best (✉)
Nelson Mandela University, Port Elizabeth, South Africa

M. Struwig
Department of Business Management, Nelson Mandela University, Port Elizabeth, South Africa
e-mail: miemie.struwig@mandela.ac.za

S. D. Carter (ed.), *COVID-19, Supply Chain, Climate Change, and Sustainable Development in Africa*, Springer Proceedings in Business and Economics,
https://doi.org/10.1007/978-3-031-26121-3_5

The United Nations Guidelines for Consumer Protection respond to this apparent contradiction by setting the promotion of sustainable consumption as one of the eight objectives of the Guidelines (United Nations Guidelines, 2003). This research aims to propose a policy framework to include sustainability in consumer protection policy.

5.2 The Global Context of Sustainability

In order to identify ways in which sustainability can be included in consumer protection policy, it is necessary to consider how sustainability entered the international policy arena. The starting point is the global debates about the respective definitions of sustainable development and sustainable consumption and production, which lay the basis for arriving at the broader and more inclusive concept of sustainability.

The 2021 annual United Nations Climate Change Conference (COP26) demonstrates how, over time, climate change has shifted from being a fringe issue to a global priority. Of particular significance is the damaging impact of climate change on lives, livelihoods, and ecosystems, and the resultant implications for production and consumption – and thus consumers.

COP26 is a prominent lodestar of sustainability, in a span of global debates that have been the focus of sustainability-related multilateral discussions and negotiations among government, business, and civil society role-players. Tracking backward, world leaders initially grappled with sustainable development, with the notion of sustainable consumption arising as a corollary. This was framed as a component of sustainable development to enable the achievement of a more sustainable world. More recently, the umbrella term of sustainability came to be used to encapsulate and cut across the concepts of sustainable development; sustainable consumption; and sustainable production more holistically and interconnectedly. The emergence of these policy concepts, within the global policy agenda on sustainability, is expanded upon below.

5.2.1 Global Sustainability Policy Agenda

The debate about sustainable development and sustainability was placed formally on the global agenda with the passing of Resolution 38/161 by the United Nations General Assembly in 1983 (Resolution 38/161, 1983), which recognized the ever-increasing deterioration of the global human environment and the concomitant strain on the availability of natural resources. The Resolution, for the first time, made explicit the perspective that for economic development initiatives to have a lasting and sustainable impact, they needed to "*take account of the interrelationships between people, resources, the environment and development*" (Resolution 38/161, 1983). Flowing from Resolution 38/161, the United Nations set up the

World Commission on Environment and Development (WCED), under the chairpersonship of Gro Harlem Brundtland, the former Prime Minister of Norway. The Commission released a report, *Our Common Future*, also known as the *Brundtland Commission Report*, in October 1987 (Brundtland Commission Report, 1987). The proposals made by the Commission are regarded as a cornerstone in arriving at a global definition of the terms sustainability and sustainable development.

The Brundtland Commission Report (1987) also triggered global policy development initiatives, facilitated by the United Nations and its various agencies. These initiatives were set up to firstly engage Governments around the world to recognize the importance of sustainability and to participate in global processes to define this concept. Secondly, the global processes aimed to encourage Governments to participate in efforts to continuously refine and develop a deeper and broader understanding of sustainability. Thirdly, and most importantly, the intention was to propose that respective Governments consider using the international definitions of sustainability and sustainable consumption as the basis for the formulation of domestic policy.

5.2.2 *Sustainable Development*

Sustainable development as a notion emerged out of what was initially a concern for the environment, arising from the fact that economic development and wealth creation were often pursued in ways that impacted negatively on the environment. This negative impact on the environment initiated a debate on how economic development could continue but be undertaken more sustainably to ensure and preserve human existence and attain more socially equitable outcomes. The Brundtland Commission Report (1987) foregrounded the sustainability challenges facing the world, calling for economic growth to be pursued in ways that would be both socially and environmentally sustainable. The Commission coined the phrase sustainable development, continuing to define sustainable development as "*development which meets the needs of the present without compromising the ability of future generations to meet their own needs*" (Brundtland Commission Report, 1987, p. 30).

Sustainable development calls for convergence between three different phenomena, namely, economic development, social equity, and environmental protection (Drexhage & Murphy, 2010). Sustainable development is thus framed as the nexus between strong economic and social development, in particular for less-resourced people, while underlining the importance of protecting the natural resource base and the environment. At the center is the imperative that economic and social well-being should not be improved in ways that destroy the environment (Dresner, 2002; Lehtonen, 2004; Robinson, 2004; Thiele, 2013). Sustainable development has been regarded as the critical means to achieving a more sustainable world.

5.2.3 Sustainable Production and Consumption

During the processes facilitated by the United Nations to develop a global consensus on the definition of sustainable development and the need to pursue this collectively, understanding started to be forged that the level of consumer consumption is a critical dimension of sustainable development. Sustainable consumption rests on the premise that patterns of consumption impact the natural resource baseline. By adjusting consumption behavior in favor of more sustainable choices, consumers can contribute to a better natural and social environment (Karsten & Reisch, 2008). From these conceptual roots within sustainable development, sustainable consumption started to come into its own and take its place on the international policy agenda in the early 1990s (Berg, 2011). Before this, only the environmental impact of the production cycle had been considered for sustainability. The impact of consumption choices had not been brought into the sustainability equation. Yet, as argued by Berg (2011), sustainable production and consumption as cojoined and interlinked processes are essentially two sides of the same coin.

The term "sustainable consumption" is subject to many interpretations. The key element of the concept of sustainable consumption and production is intergenerational, and current consumption patterns are tailored to meet the needs of present generations without compromising the needs of future ones. In addition, sustainable consumption incorporates social considerations, responding to basic needs and quality of life, doing so in resource-efficient ways, and minimizing the use of natural resources.

Sustainable consumption starts with sustainable production methods. Sustainable production is the creation of goods and services using processes and systems that are nonpolluting, conserving energy and natural resources; economically efficient; safe, and not health-harming for workers, communities, and consumers socially, while also being creatively rewarding for people in production. Sustainable production needs to consider the entire life cycle of the service or product. Initiatives toward more sustainable production need to be systems-wide and encompass the full spectrum of each component within the value chain of production and consumption.

Sustainable consumption is not a new concept. It is one of the eight objectives of the United Nations Guidelines for Consumer Protection, drawn up in 1999. These Guidelines (United Nations Guidelines, 2003) are quite explicit, stating the rights of consumers upfront and indicating that "*consumers should have the right to promote just, equitable and sustainable economic and social development and environmental protection*." Global policy frameworks thus recognize sustainable consumption as an important behavior change mechanism that could additionally contribute to the achievement of sustainability.

5.2.4 Responsible Consumption

The notion of responsible consumption sits alongside sustainable consumption. Responsible consumption is about changing consumption patterns using a three-pronged approach, firstly, by consuming less, secondly, by becoming more resource-conscientious and thus making better consumption choices, and, lastly, by consuming more efficiently. Responsible consumption requires consumers to pursue changes proactively in their consumption patterns, with the over-arching aim being that current and future generations consume within the socio-economic and environmental constraints of the earth's finite resources.

Consumers are key drivers of sustainability through their consumer choices. Consumers thus need to be persuaded to change their behavior and adopt more sustainable patterns of consumption.

5.2.5 Sustainability

The debates about the respective definitions of sustainable development and responsible, sustainable consumption, and production lay the basis for the broader and more inclusive concept of sustainability. Owing to the extensiveness and multifaceted nature of sustainability, arriving at a single definition of sustainability is a challenge. A wide range of definitions exists in literature, as many disciplines have provided a definition, including ecologists, economists, sociologists, environmentalists, and, more recently, business management scholars (Faber et al., 2005). The literature also suggests that most definitions include a recognition of the connection between economic systems operating within environmental resource constraints, taking cognizance of the quality of life of present and future generations (Smith, 2014). In addition, sustainability affects society as a whole and is linked to human survival, with sustainability thus defined as the ability to continue a defined environmental, economic, and social activity indefinitely (Sustainability Root Cause, 2016). Strong (2001) argues that the goal of sustainability must be to improve the lives and prospects of people environmentally and socially as well as economically. Sustainability refers to the coordination of economic, social, and environmental issues in support of the long-term viability of people and the planet.

To define sustainability, the first important dimension to note is the confluence and interlinked connectivity of economic, social, and environmental considerations. The second dimension to note is the usage of global resources in ways that meet basic needs yet with due concern for the needs of future generations. The third dimension of sustainability is that it is multifaceted and needs to be considered at all stages of the goods and services production as well as the consumption cycle. Considering these three fundamental dimensions, for this research, sustainability is defined as "*an integrated and holistic approach to production and consumption which inter-links economic, social and environmental considerations and is mindful of the impact on global resources; takes future generations needs into consideration; and is multi-faceted, straddling all dimensions of the goods and services production and consumption life-cycle*."

5.2.5.1 Sustainable Consumption and Consumer Protection Policy

The link between sustainable consumption and consumer protection policy was first proposed on a global level in 1995 when the UN Economic and Social Council (ECOSOC) requested the incorporation of sustainable consumption into the UN Guidelines on Consumer Protection. This was consolidated as a policy proposition in 1999, when the Guidelines were amended accordingly (United Nations Guidelines, 2003), to include a section with several provisions dealing with sustainable consumption. A distinctive and notably significant global policy-setting initiative impacting sustainability, sustainable consumption, and consumer protection which has happened after the 2012 Rio + 20 or Earth Summit 2012, and the third International Conference on Sustainable Development, is the adoption of the Sustainable Development Goals [SDGs]. These 17 goals were adopted in September 2015 by the United Nations General Assembly and are incorporated within the 2030 Agenda for Sustainable Development. Consumer policy is pivotal in the achievement of SDG 12 in particular. SDG 12 is to "*ensure sustainable consumption and production patterns*." A consumer protection policy can be an effective tool to ensure that consumers have access to accurate and easy-to-understand information about product sustainability. This, in turn, will enable consumers to understand how their consumption choices can impact the environment and make informed choices in this regard. The policy can also be designed to put measures in place to improve corporate management of sustainability and its reporting. Beyond, and in addition to this, consumer policy is an important mechanism to support the implementation of other, if not all, of the 17 SDGs. The amended United Nations Guidelines for Consumer Protection, as adopted by the General Assembly in its resolution 70/186 of 22 December 2015, are the most relevant instrument for this purpose. Incorporating sustainable consumption into consumer protection policy brings to the fore the formidable potential of this public policy as a mechanism for achieving economic growth and inclusive and sustainable development. Consumer protection policy is central to building a fairer, safer world (Consumers International, 2015), and with growing global concern for sustainability (Smith, 2014), the incorporation of sustainability into consumer protection policy is critical. It is, however, important to take cognizance that developing countries may lack consumer policy frameworks or may have too recently implemented such frameworks and/or may face challenges about consumer protection law enforcement. Intergovernmental support, policy advice, and technical assistance for developing countries and countries with economies in transition would be an important element for the achievement of the 2030 Agenda and the SDGs – particularly SDG 12.

5.3 Stakeholder Participation

Effective consumer protection requires the active participation of a range of stakeholders, who play different roles but need to do so in an integrated and coordinated manner. Vila (2017) suggests that states are increasingly assuming regulation and surveillance while transactional market players, namely, businesses and consumers (either as individuals or in consumer formations), should play respective roles in enabling compliance and ensuring good business practices, complemented by consumer education empowerment strategies. Stakeholder engagement, agency, advocacy, and voice are key to the successful development and implementation of consumer protection policy and related regulatory frameworks. The primary stakeholders are government; business; and consumers.

5.3.1 Governments

Public authorities at the local, national, regional, and global levels can influence the sustainability of consumption and production by providing a framework within which businesses and consumers operate and transact. Policy frameworks can include provisions that direct businesses and industry to develop self-regulatory measures implemented through sector ombuds. Frameworks can also have provisions that set out mandatory compliance obligations for producers, with concomitant adherence and enforcement tools such as taxes on unsustainable goods and services. Soft law, including best practices, codes of conduct and guidelines for business, and civil society initiatives targeting consumers, are necessary instruments that complement compliance regimes. Governments can also provide enabling conditions for consumer education and information campaigns that encourage businesses and consumers to adopt sustainable consumption patterns.

Poorly designed state regulatory frameworks can be burdensome to implement and result in excessive compliance costs, which, in turn, may become the cause of noncompliance with the policy. The appropriateness of compliance measures and sanctions for noncompliance also need to be carefully thought through, in consultation with those affected by these, as well as those who will be responsible for complying with and implementing policy imperatives – in this case, businesses and consumers. As a result, the government must invite business and consumer representatives into state-led policy development initiatives.

The multifaceted nature of sustainable consumption can result in responsibility for policy development and implementation being vested in more than one level of government and spanning across several line function departments at a national level. The impact of this fragmentation is that important dimensions of sustainability can be overlooked due to policy lacunae and implementation disconnects. Policy coordination mechanisms between government departments need to be devised to enable horizontal and vertical integration, for example, through establishing a sustainable consumption cluster or hub consisting of all departments and entities with responsibility for a component of sustainable consumption. Increased policy coordination would help to ensure that, as consumer protection is encouraged and promoted, sustainable consumption is also encouraged and practiced at all levels of society. A collaborative intra- and intergovernment effort is required to serve as an effective means of promoting sustainability.

Governments would do well to consider the sustainable consumption provisions in the United Nations Guidelines on Consumer Protection as at least a preliminary basis for crafting and strengthening consumer protection laws and policies. This, in turn, will contribute to the development of mechanisms to enable governments to better fulfill their commitments of Agenda 2030, and thus the SDGs. Consumer protection and sustainable consumption policy need to integrate horizontally across government and connect to environmental and social policy as well as economic growth and development strategies.

5.3.2 Business

Business and industry role-players are at the center of implementing consumer protection policy. Business cooperation in implementing consumer protection, sustainable consumption, and sustainable production modes and models is essential to efficacy and meaningful impact. When introducing new policy measures, doing so in a manner that requires incremental compliance may serve to reduce possible resistance to compliance, while incentives can encourage businesses to make positive changes toward achieving compliance. Sectoral self-regulation is another avenue to encourage the adoption of and participation in sustainability measures, as this allows business role-players to take the lead in designing and developing sustainability frameworks for the relevant sector. This also creates the impetus to set up customized alternative dispute resolution mechanisms such as ombuds, to address sector-specific consumer disputes and provide redress.

The role of business is also pivotal in sustainable production. To achieve this, in many instances, a business will need to adopt practices and procedures at all stages of the production life cycle to respond to the incorporation of sustainability into consumer protection policies (Mont & Bleischwitz, 2007). Using a life-cycle approach to examine the economic, social, and environmental impacts of goods and services at each stage of the production process, from product design through to disposal, and then proactively adapting, innovating, and (re)designing production processes to be more sustainable, is taking root in global corporates. Public policy is also increasingly making this an imperative.

Trends are beginning to show that the formal, big business context is shifting toward a more holistic conceptualization of sustainability. Specifically, sustainability is recognized as a value that needs to be deliberately and consciously built into the design and essence of a business as part of the strategic focus of the business. Doing so has come to be regarded as good corporate citizenship while simultaneously offering a potential competitive advantage. Sustainability should also be explored as a means to reduce the (medium/long term) costs of doing business and positively differentiate products in the market. This would have the knock-on effect of increasing demand by attracting new customers and building loyalty and pride in existing customers. Growing and strengthening consumer consciousness around global warming and the impact of climate change is a strong push factor in this regard. Global forums, such as COP26, have also produced resolutions that commit governments to sustainability-linked targets, which would need to see a trickle-down to reflect in national sustainability and consumer protection policy and sector regulation.

5.3.3 Consumers and Consumer Groups

Consumers, through their purchasing power, are at the heart of the economy and hence play an important role in a country's economic growth and development. Consumer trust is crucial for the expansion of economic activities. Consumers need to be empowered and encouraged to make informed, sustainable, and healthy choices. Consumer education is the bedrock of consumer empowerment. Having access to product information enables consumers to become increasingly aware of the environmental impact of their purchases, which in turn can influence consumers toward exercising choices that lead to more sustainable consumption behavior. Including sustainability into consumer protection policy can shift the consumption patterns of consumers in ways that promote sustainable lifestyles; avoid irreversible depletion of resources; reduce pressure on the environment; save money; and provide a better quality of life.

Consumers and consumer organizations have a critical role to play in lobbying the government and advocating for consumer policy where this is not yet in place. Consumer advocacy must engage with the government, extending to campaigning as may be necessary to ensure that sustainable consumption is incorporated into policy. Consumers also have a fundamental role to play in ensuring adherence by businesses to consumer protection regulatory frameworks. Consumers must actively demand redress when obligations have not been fulfilled and consumer rights have been trampled upon by businesses and the state. Consumer pressure can also convince businesses to change production and supply methods toward more sustainable modes. The most direct actions would be to create public awareness of noncompliance and ultimately withdraw consumer purchasing power. Global consumer advocacy for sustainability by consumer organizations, alliances, and campaigns has seen significant gains. The outcome of advocacy such as that of Consumers International (Consumers International, 2015) and the growing global awareness of sustainability issues have seen several international agencies, and inter-country bodies such as the United Nations, the Organisation for Economic Co-operation and Development, and the European Union develop guidelines for incorporating sustainability within consumer policies. Consumer advocacy also contributed to the adoption of SDG 12.

5.3.4 Multi-Stakeholder Global Governance

With international trade becoming increasingly cross-border, and as e-commerce and the digital economy are embraced by consumers, it has become essential for governments to collaborate to establish inter-country cooperation in pursuit of consumer protection and sustainable consumption. The United Nations Conference on Trade and Development (UNCTAD) has played and continues to play a pivotal role in offering a platform for governments to engage in the ways to establish international institutional machinery to promote cooperation. UNCTAD has the responsibility for facilitating the Intergovernmental Group of Experts on Consumer Protection Law and Policy (IGE), which is, in turn, mandated to oversee the implementation of the United Nations Guidelines on Consumer Protection. The IGE offers a platform for promoting and strengthening international multilateralist governance on consumer protection matters, with a particular focus on sustainable consumption as set out in the United Nations Guidelines on Consumer Protection. The IGE is open to broad participation by governments; business and trade interest groups; consumer groups; and academia. The IGE is emerging as a functional global space for multi-stakeholder engagement and participation in the international institutional machinery for consumer protection. It offers a platform for continuously exploring ways to promote sustainable consumption within an expanding global consumer protection agenda (Vila, 2017]. This public policy domain is now a cross-cutting dimension that is embodied within the SDGs. Recognizing that consumer policy can be an effective instrument to promote sustainable consumption, it is now being considered by several organizations and institutions in the United Nations system.

Given the enormity of the challenges about sustainable development, sustainable production and consumption, and responsible consumption, finding workable solutions requires engagement and contributions from all stakeholders such as governments, business, consumers and their organizations, and international bodies. The UNCTAD IGE is proving to be an important platform in this regard, fostering cooperation and creative partnerships to progressively build toward the Global 2030 Agenda and the achievement of the SDGs.

5.4 Qualitative Research

The qualitative research was part of a research thesis that was supported by the Research Partnership Platform of UNCTAD, which is aligned with and feeds into the work of the IGE. In the study, document analysis using content analysis was used in instances where the UNCTAD member states provided copies of legislation and policy documents and responded to open-ended questionnaires that were distributed. A content analysis schedule was developed to define the categories or themes of content that were to be considered as the topics of interest within the documents received. The words and themes for the content analysis were extracted from the United Nations Guidelines on Consumer Protection (2003) which deal specifically with sustainability and sustainable consumption.

The research proposition, based on an extensive literature review, proposed four components for a theoretical framework to include sustainability in consumer protection, namely:

- Sustainability definition elements
- Product life-cycle model
- Sector self-regulation for sustainability
- Policy instruments

The policy frameworks and legislation of countries in the sample, namely, Egypt, Botswana, Zambia, Mauritius, Tanzania, and South Africa, were examined to ascertain which of the four components were contained in the respective policy and legislative provisions. Novel policy provisions, which included sustainable consumption in consumer protection policy, and in-country regulatory frameworks, were also extracted and recorded where these were identified.

Official state documents were the primary source of data for this study. Written public documents such as laws that have been passed by the legislative authority of a country were examined. In African countries, where the use of the Internet as a government communication tool is still gaining ground, it was difficult to access these documents. Direct requests for copies of legislative acts and policy documents were made from the purposive sample. Twelve United Nations Conference on Trade and Development (UNCTAD) member states in Africa were included in the purposive sample. These countries were selected on the recommendation of expert officials at UNCTAD. Only six out of the 12 (50%) member countries responded to the request for information.

5.4.1 The Results of the Qualitative Study

The results of this qualitative inquiry determining how a range of countries have included sustainability in consumer protection policy showed that all the countries examined had at least some elements of environmental considerations in their legislation but to a lesser extent in African countries with less-developed economies. African countries tended to focus on more narrow environmental policy implications than on the more broadly defined notion of sustainability. Indigenous consumption systems, which by their nature are very sustainable, were not explored in the study. Policy mechanisms that promote sustainability were more evident in the policies and laws of developed countries. Developed countries have begun to regard sustainable household consumption measures as an important policy route for sustainability in general, and certain aspects of consumer protection. Including sustainability in consumer protection remained complex policy terrain, and there was a need to set out the nexus between sustainability and consumer protection in any policy exposition. In the case of most African countries, basic consumption needs were foregrounded as the primary concerns of consumers, ahead of sustainability concerns. Further, poverty limited consumer choices, particularly if more sustainably produced and eco-efficient goods came at a higher price. Interpretive observations regarding policy measures that have enabled the inclusion of sustainability into consumer protection also emerged from the qualitative inquiry and are set out in Table 5.1.

The interpretive observations in Table 5.1 provide guidance as to the actual policy contents that should be included under each of the four elements of the theoretical framework so that the theoretical framework translates into implementability. This is shown in Fig. 5.1.

Table 5.1 Interpretive observations from the content analysis

Definition elements	Sustainability needed to be stated as a policy intention and the link between consumer protection and sustainability needed to be made explicit in the policy proposition Sustainability needed to be clearly defined in the policy, encapsulating the multifaceted dimensions of sustainability as a concept, because sustainability is conceptually more integrative and multidimensional than purely environmental concerns
Product life cycle	Sustainable production rested on specific initiatives to be implemented at each stage of the production and consumption cycle Sustainable production was more achievable if responsibility for this was apportioned across all role-players in the cycle
Sector self-regulation	Policy should prescribe a minimum mandatory compliance requirement to include sustainability in all sector initiatives, complemented by voluntary sectoral self-regulation Introducing sustainability measures into the business environment should be approached through incremental and voluntary adaptation of business practices
Policy measures	A mix of policy instruments could be developed to include sustainability, starting with an expanded approach to product safety Product information policy mechanisms and consumer information standards enabled consumers to make more sustainable consumption choices Sustainability straddled multiple policy domains and effective policy implementation required coordination and cooperation and joined-up implementation across government departments and spheres or levels of government Benchmarking against international guidelines and best practices was useful to inform domestic consumer protection policy
Contextual considerations	Economic conditions in poorer countries necessitated placing the basic needs of consumers ahead of sustainability considerations Individual consumers could impact sustainability through their choices and behavior and thus consumer education and stakeholder engagement were important and created an enabling environment for policy implementation Policy initiatives and incentives to promote sustainable household consumption should be developed as a means to adjust individual resource consumption patterns

Source: Researcher's Construct

Figure 5.1 shows that the four aspects of the theoretical framework were found to be relevant and should hence be retained as the basis for developing the framework for including sustainability in consumer protection policy. Several countries were found to have recently amended their consumer protection legislation, elements of which were more inclusive of sustainability. Other countries, and African countries, in particular, indicated that policy reviews were taking place or were imminent. This provides an opportunity for these countries to amend the policy to include sustainability measures customized for the particular context of each country.

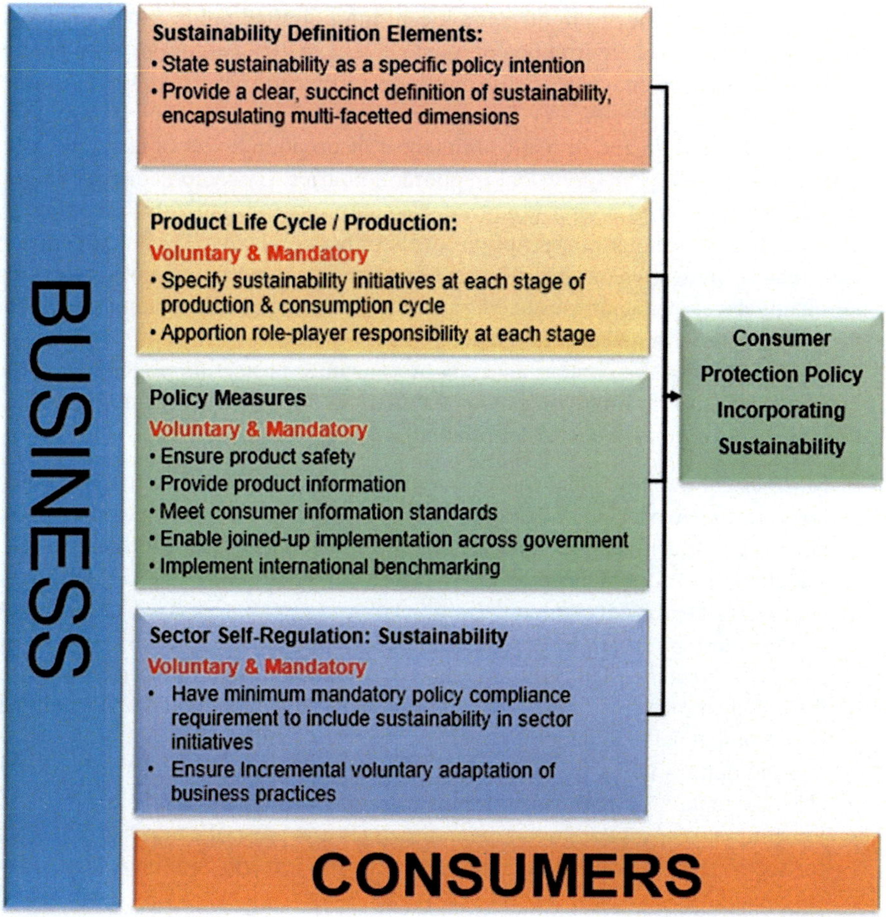

Fig. 5.1 Proposed policy framework to include sustainability. (Source: Researcher's Construct)

5.5 Conclusions and Recommendations

The results of the content analysis and the responses to the open-ended questionnaires indicated that economically more-developed African countries had put in place a suite of measures to include sustainability in consumer protection policy and to enable consumers to make more sustainable choices. These ranged from product safety regulations to product eco-labeling to provide sustainability-related information such as genetically modified product ingredients, and energy usage of the product. Countries had also established regulatory bodies such as a Consumer Commission to enable the implementation of policy and to ensure compliance with the regulatory requirements by businesses – at least those operating in the formal sector.

Most African countries in the sample had more limited policy measures related to product safety and environmental management. African countries may be pressured by immediate budgetary constraints to prioritize the basic needs of consumers such as education, healthcare, and transport, ahead of sustainable consumption measures. Poverty in and of itself limits the consumption levels of consumers, and subsistence lifestyles are prevalent in poorer countries. By socio-economic circumstances, consumers in poorer countries consume more sustainably in comparison with the high consumer consumption levels in economically developed countries. This may be indicative of why the consumer protection policy regimes in less economically developed countries had less explicit measures incorporating sustainability within policy frameworks and legislation.

The findings of the content analysis of consumer protection policies in certain countries point to the following recommendations for mechanisms to include sustainability provisions in consumer protection policy:

- Make use of the functional role of consumer protection agencies to ensure ongoing policy development as an opportunity to research and introduce policy mechanisms that specifically include sustainable consumption provisions and speak to sustainability more inclusively.
- Expanding and/or developing definitions of *unconscionable conduct*; prohibited *conduct* and *unfair trading*, so that these extend to include unsustainable consumption practices. This could also be developed through jurisprudence during the adjudication of matters by tribunals and courts, which would, in turn, impact future policy development by government departments.
- Expand definitions of product safety to extend to products produced unsustainably and make this applicable at each stage of the product life cycle.
- Issue ministerial and/or government directives that prescribe minimum standards for sector self-regulation which must specifically include minimum standards on sustainability.
- Invite stakeholder involvement, notably business and consumer organizations, to participate in policy formulation while creating enabling conditions for stakeholder advocacy, particularly as this relates to the incorporation of sustainability provisions in consumer protection policies.

- Ensure that stakeholder education in general, and consumer education in particular, is inclusive of how to achieve more sustainable consumption at the level of the individual consumer; in households; in business processes throughout the product life; and in government practices.

These policy advances offer useful opportunities to include sustainability in consumer protection policy regimes. The policy review processes that were indicated to be underway in several countries in the study also provide an opportunity to include legislative provisions that deal with sustainability.

In conclusion, consumer protection policy stakeholders, namely, governments, businesses, and consumer organizations, should refer to guidelines on consumer protection, developed by global organizations such as the United Nations and the Organisation for Economic Co-operation and Development (OECD) to strengthen and improve policy provisions to include sustainable consumption in consumer protection policy. This will contribute to the global effort to achieve the sustainability agenda of 2030 as set out in the 17 Sustainable Development Goals, specifically Goal 12: to "*ensure sustainable consumption and production patterns.*"

This study also brought to the fore future studies that can be undertaken to explore the research question about the relation and link between sustainability imperatives and consumer choices, disaggregated by class and educational levels. Exploring the impact of factors such as access to information and technology on consumers' attitudes toward sustainable consumption practices is another area of future research. The lack of Afrocentric scholarship in public policy theorization and the need for more research about how consumer protection is resultantly affected, as well as that consumer behavior in marginal economies and informal markets, should also be the subjects of future research studies. The growth of unbordered global e-commerce points to the need for research on the impact of this on sustainability across the entire product life cycle, including e-waste generation and individualized product delivery that characterizes online shopping.

Acknowledgments This work has been extracted from the PhD of Laura Best titled: A Framework to Incorporate Sustainability into South African Consumer Protection Policy.

References

Berg, A. G. (2011). *Inequality & Unsustainable Growth: Two sides of the same coin? IMF staff discussion note 11/08* (Vol. 11, p. 1). IMF.

Brundtland Commission Report. (1987). *Report of the world commission on environment and development: Our common future, from one earth to one world.* http://www.un-documents.net/our-common-future.pdf

Consumers International. (2015). *Commitment to sustainability.* http://www.consumersinternational.org/who-we-are/ci-and-sustainablity/

Dresner, S. (2002). *The principles of sustainability.* Earthscan.

Drexhage, J., & Murphy, D. (2010). *Sustainable development from Brundtland to Rio 2010.* United Nations.

Faber, N., Jorna, R., & Van Engelen, J. (2005). The sustainability of sustainability – A study into the conceptual framework of the notion of sustainability. *Journal of Environmental Assessment Policy and Management, 7*(1), 1–33.

Gumede, V. (2008). Public policy making in a post-apartheid South Africa: A preliminary perspective. *Africa, 38*(2), 7–23.

Karsten, J., & Reisch, L. A. (2008). Sustainability policy and the law. *German Policy Studies, 4*(1), 45–66.

Lehtonen, M. (2004). The environmental-social Interface of sustainable development: Capabilities, social capital, institutions. *Ecological Economics, 49*, 199–214.

Mont, O., & Bleischwitz, R. (2007). Sustainable consumption and resources Management in the Light of life cycle thinking. *Environmental Policy and Governance, 17*(1), 59–76.

Resolution 38/161. (1983). *United Nations general assembly: Process of preparation of the environmental perspective to the Year 2000 and beyond.* http://www.un.org/documents/ga/res/38/a38r161.htm

Robinson, J. (2004). Squaring the circle? Some thoughts on the idea of sustainable development. *Ecological Economics, 48*, 369–384.

Smith, E. E. (2014). Beyond greening: Reflections on the business sustainability imperative, Professorial Inaugural Lecture delivered, Nelson Mandela Metropolitan University.

Strong, M. (2001). *Where on earth are we going? Alfred a.* Knopf.

Sustainability Root Cause. (2016). *Finding and resolving the root causes of the sustainability problem.* http://www.thwink.org/sustain/glossary/Sustainability.htm

Thiele, L. P. (2013). *Sustainability.* Polity Press.

United Nations Guidelines. (2003). *United Nations guidelines for consumer protection.* United Nations.

Vila, I. (2017). The legal and institutional framework of consumer protection. 2nd session, 3 and 4 July United Nations conference on trade and development.

Waller, P., Morris, R.M., & Simpson, D. (2014). *Understanding the formulation and development of government policy in the context of FOI.* http://ico.org.uk/about_us/research/~/media/documents/library/Freedom_of_Information/Research_and_reports/UCL_REPORT_GOVERNMENT_POLICY_IN_THE_CONTEXT_OF_FOI.ashx

Chapter 6
Increase Business Performance Through the Use of Virtual Technology

Ioan David Legman

6.1 Introduction

The current global economic situation requires certain changes at firm level. In addition to the managerial competence which is a defining element of the company's strategy, innovation plays an increasingly important role in the business. Innovation can be felt differently across the firm, for some employees it requires a state of tension and discomfort, and for others it can be a chance to make it easier for them to do their job.

Sometimes innovation brings with it new technologies. The way in which these technologies are implemented can be decisive in the success of the firm's market. Employees will use the technology adopted in an organization anyway, but the success of the implementation depends on a high level of acceptance that reflects a voluntary use based on intrinsic motivation. What is being sought today regarding the workplace is a collaborative management; team-related skills can be a strong point when choosing a candidate for a position in a company. The latest research has concluded that there is an underutilized performance lever in terms of teamwork.

The introduction of new indicators to measure employee performance was not sufficient to drive economic growth within a firm. These elements must be combined so that employees feel they can really contribute to the economic growth the company is aiming for. Thus, in addition to increased productivity, reduced costs, economic essentials, high customer and employee satisfaction are also needed.

The fundamental objective of this work is to identify and analyze how to improve the performance of some firms in terms of ensuring a functional relationship between the SIM method and the Spc indicator.

I. D. Legman (✉)
University of Medicine, Pharmacy, Sciences and Technology of Targu Mures, Targu-Mures, Romania

S. D. Carter (ed.), *COVID-19, Supply Chain, Climate Change, and Sustainable Development in Africa*, Springer Proceedings in Business and Economics,
https://doi.org/10.1007/978-3-031-26121-3_6

The central assumption derives from previous studies over several years, which were based on the assumption that in economic activity the basic idea continues to be maximization of profit. On the basis of this assumption, it was assumed that the economic activity in addition to obtaining the profit must be based on the satisfaction of the customer, but also take into account the performance of the employees.

The image and reputation of a company are intangible assets whose importance is growing. This identity is directly associated with the personality and self-operation of the company, which in turn implements its "DNA" on the employee who comes into contact with the customer.

6.2 Material and Methods

Two companies from different countries were included in this study. These companies were from different sectors of activity, so the postal services company participated in the study with 67 employees, offering delivery services packages and envelopes, while the second company participated in the study with 6 employees offering hair-style services to customers.

The basic indicator used in this study was the Spc indicator, which, depending on the area in which it was applied, has undergone some adjustments. If in the online trade the interaction between the employee and the customer was based on a telephone call, chat, or email, in the case of the courier company and the hair-style lounge, the interaction was only physical. One thing is sure the customer satisfaction regardless of how the interaction should be maximum. The research has been carried out over several years, but in the present research on the courier company and the hair-style lounge has been carried out for 3 months.

Within the courier service company, the map (weighted arithmetic average of customer satisfaction) was applied by email sent throughout the day to the customers, in which they expressed their satisfaction/dissatisfaction as a result of their interaction with the employee of the company. This element shows us the degree of satisfaction with regard to the quality of transport, the quality of communication with employees, the promise of the employees from the customer's perspective is assessed, but the customer's opinion on the company's prices on the market will also be taken into account. In the case of the hair styling map lounge, the questionnaire was applied and sent to the customer on the WhatsApp, which was written in Google Forms.

An employee of the courier company delivers on average about 60 parcels per day of different sizes. The company's employees' program is flexible. It ranges from 7 hours/day to 12 hours/day. There are situations where employees, for some reason, have been out of work. The other items used to calculate the Spc indicator are t (weighted average unit delivery time), WmL (average labor productivity), and Σc (sum of delivered packages).

According to Table 6.1 of the program, 67 employees were included from the courier company, and 6 employees were included in the parlor of the hair style.

Table 6.1 Average values for the SPC indicator variables during the implementation period

Month	Contacts		Worked hours	Cost/hour/employee (euro)	Maps	t	WmL	Σc	Σc/h	Average value of Spc and interpretation	
	Chat	Phone									
POSTAL services (67 employees)											
July 2020	–	–	138.0	3.25	3.38	6.86	8.98	1206	9.00	0.105	**Medium**
August 2020	–	–	155.7	3.25	3.45	6.60	9.20	1422	9.20	0.132	**Medium**
September 2020	–	–	162.3	3.25	3.27	6.40	9.47	1535	9.46	0.140	**Medium**
Hair-style SALON(6 employees)											
Aug 2021	–	–	157.6	25	3.03	0.61	1.63	256.5	1.63	0.083	**Weak**
Sep 2021	–	–	161.8	25	3.17	0.61	1.65	266.5	1.65	0.093	**Weak**
Oct 2021	–	–	166.0	25	3.72	0.56	1.79	296.5	1.79	0.144	**Medium**

Source: Data collected from companies

As regards the activities of the employees in the hair-style salon, they served on average 1.63 customers per hour, as shown in Table 6.1 in August, with a slight increase in September to 1.65 customers per hour, and in October, it increased by 9% from August to 1.79 customers per hour.

With regard to employee productivity by comparing the results of the two companies, an increase can be observed in relation to the base month. In the case of the courier company, we have an increase in September 2020 by 5.45% compared to July 2020; while in the case of the salon, we have an increase in productivity of around 10% in October compared to August 2021.

By comparing the Spc indicator in the two companies, we can see in Fig. 6.1 that the trend is for growth, for improving the indicator.

Continuous innovation is needed to meet customers' needs, respond to their needs, and maintain a high degree of competitiveness. Analyzing the program implemented, it was considered that employees were having difficulty in putting packages properly into the utilities they used. These difficulties were also encountered in the warehouse where there were unused spaces for certain reasons.

As a result, the use of VR/AR technology in which employees had different refresher sessions was introduced to increase the efficiency of their employees in the use of storage space, i.e., customer satisfaction, labor productivity, etc. In these programs where VR technology was used, employees were able to experiment with different methods of arranging parcels in different storage areas. The arrangement of parcels has been experimented with in both classical warehouses and in various trucks (Fig. 6.2).

The Spc indicator brings a real picture within the company of the relationship of customer satisfaction – employee performance, an important element in increasing company efficiency. To increase this employee performance, it was necessary to

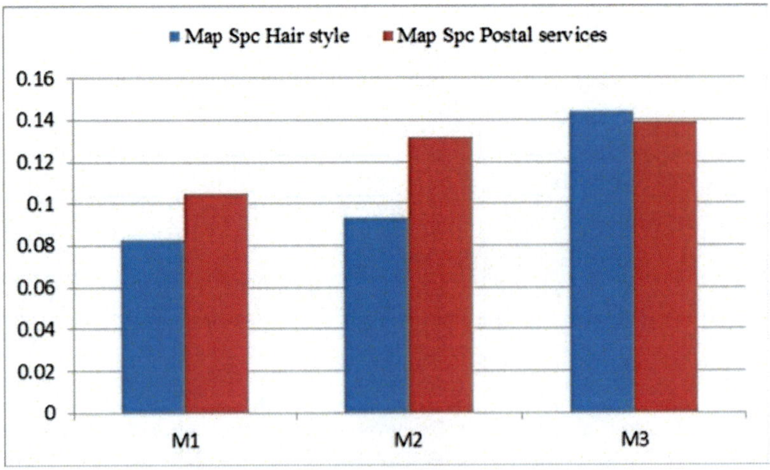

Fig. 6.1 The average of the Spc indicator – 3-month benchmarking. (Source: Data obtained from the company)

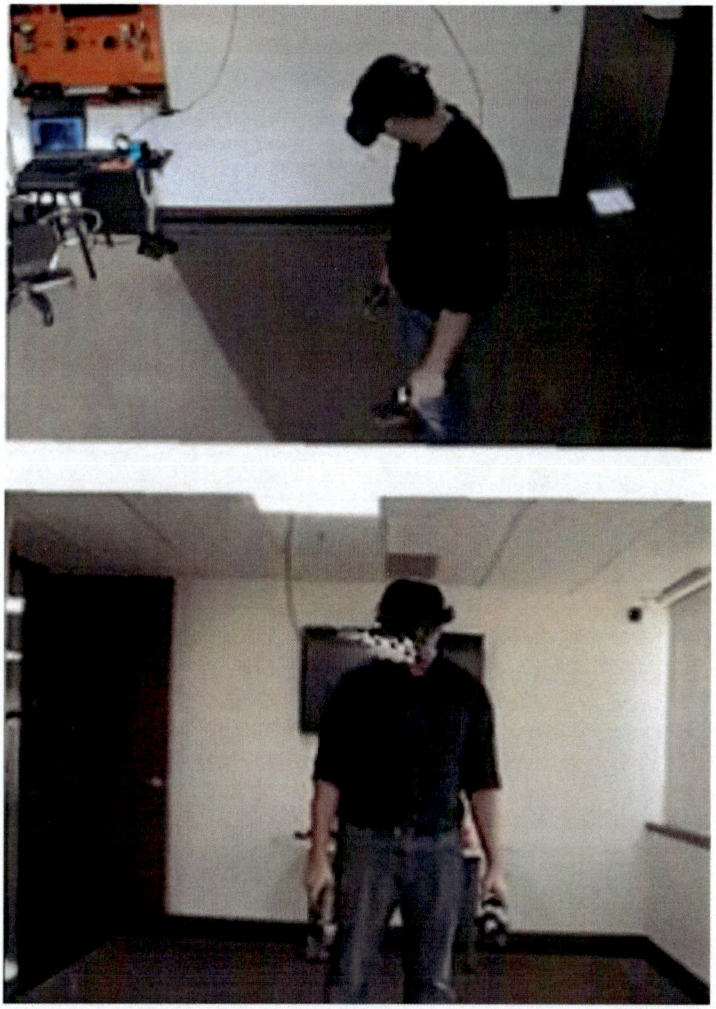

Fig. 6.2 Employee using VR technology for package arrangement. (Source: Data obtained from the firm)

introduce the SIAR program. This program consists of the Spc indicator, the ITA method, and the technology AR that is technology would use virtual world elements in real life (Legman & Gabor, 2020a, b). Employees in this way can see in advance how packages are arranged in a specific building, storage space, transport unit, etc. The use of the specialization program in virtual space (VR) is for a limited period followed by a break period. This is because virtual space causes certain health problems for users of VR devices (Filipowiak, 2020).

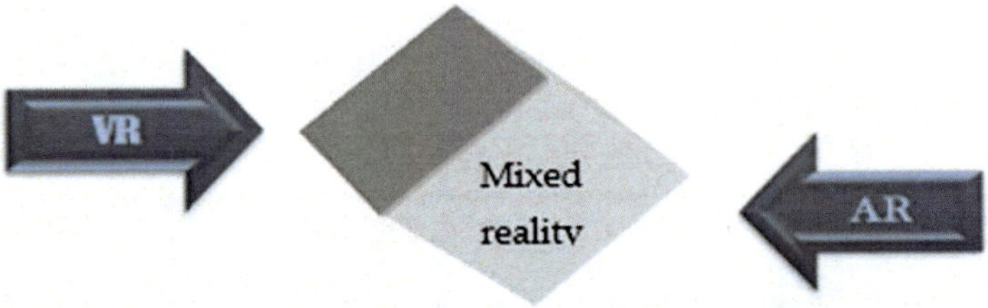

Fig. 6.3 "Virtual reality" and "augmented reality" are important elements of mixed reality. (Source: Legman, I.D., Gabor R.M. (in press) Augmented Reality technology – a sustainable element for Industry 4.0, Acta Marisiensis, Seria Oeconomica, vol. 2, p.17–36)

Fig. 6.4 How to arrange packages in VR. (Source: Data collected form company)

Fig. 6.5 Customer trying various colors to apply hair paint. (Source: Data collected form hair-style salon)

This program aimed to specialize employees on how to sort units by size and adapt to different deposit spaces; in other words the mixed reality part was experienced as well (Figs. 6.3 and 6.4).

In the hair-style lounge, the customers were able to use items from augmented reality to experience some of the possible colors. Customers were surprised that they could see which of the hair colors best suits them before performing the color-changing treatment (Fig. 6.5).

6.3 Results and Discussion

Looking at the 3 months from the perspective of the economic efficiency of the firms analyzed, we can infer that the economic activity of these companies has undergone some changes. First, labor productivity has increased visibly.

From the Spc perspective, from July to September, a trend can be observed for employees in the red sector (weak area) of the Spc indicator to the yellow area (medium area) of the Spc indicator. So if in July we had 31 employees in the weak zone, in August the number reached 9 and in September the number of employees fell to 5 (Table 6.2).

Some elements related to the creation of the Spc indicator show that the economic situation of society is on the rise. If we look at the weighted arithmetic mean of the Spc indicator, we can see a 25% increase in August as compared to July, and a 5-percentage point increase in September as compared to August.

All these elements have led to an increase in the economic efficiency of the courier company, which is also reflected in the economic data of the company. Thus, Table 6.3 shows the main economic indicators of the courier company.

As for the turnover indicator, an increase of over 2% can be observed in August as compared to July. The company's profit also increased in August as compared to July by more than 14%, the upward trend continued in September when it reached 7500 euro representing a 25% increase in September as compared to August (Fig. 6.6).

The company's expenses remained at the same ceiling in August as compared to July, with a slight decrease in September as compared to August. As regards the economic efficiency of the yield of the inputs, this is continuously increasing over the 3 months. The second indicator of economic efficiency, the one on input consumption, is on a decreasing trend.The rate of economic return is also on the rise in August from July by 13% and an increase of 26% in September from August. The economic activity of the courier company is quite stable in terms of the economic indicator profit, which is positive throughout the year (Capelle, 2014) (Fig. 6.7).

Table 6.2 Evolution of the economic indicators (July–September 2020) used for the achievement of the Spc indicator

Indicators	July	August	September
Map Spc	0.105	0.132	0.139
ΔMap Spc%		25.71	5.30
WmL	8.976	9.197	9.465
ΔWmL%		2.46	2.91
∑C	80,798	95,292	102,826
ΔC%		17.94	7.91
Hours worked/day	6.901	7.783	8.116
ΔL/day%		12.78%	4.28%

Sources: Own calculations based on information collected from the firm of postal services

Table 6.3 Evolution of economic indicators for the courier company July–September 2020

Economic indicators	July 2020	August 2020	September 2020
Turnover	44,630	45,728	47,059
ΔQP%		2.46%	2.91%
Profit	5219	5966	7500
ΔPr%		14.31%	25.71%
Total costs	39,411	39,762	39,559
ΔCT%		0.89%	−0.51%
Number of employees	88	88	87
Eerfp	1.13	1.15	1.19
Eecfp	0.88	0.87	0.84
Rate of return	13.24%	15.00%	18.96%
ΔRprCT%		13.30%	26.35%
The commercial rate	11.69%	13.05%	15.94%
ΔRprQP%		11.56%	22.15%

Sources: Own calculations based on information collected from the firm
ΔQP% – percentage change in turnover
Eerfp – economic efficiency in terms of factor yield
Eecfp – economic efficiency in input consumption
ΔRprCT% – percentage change in rate of return
ΔRprQP% – percentage change in the commercial rate

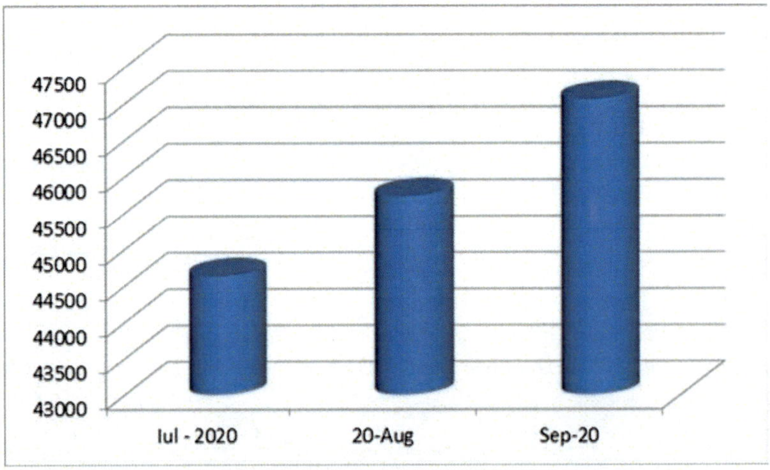

Fig. 6.6 Turnover of postal service company. (Sources: Own calculations based on information collected from the firm)

At the end of the year, the company totaled 1,071,814 packages dispatched, providing a turnover of 535,560 euro (Fig. 6.8).

Economic data show that between July and September 2020, some changes took place, with the peak of the company's turnover reaching 47,059 euro in September.

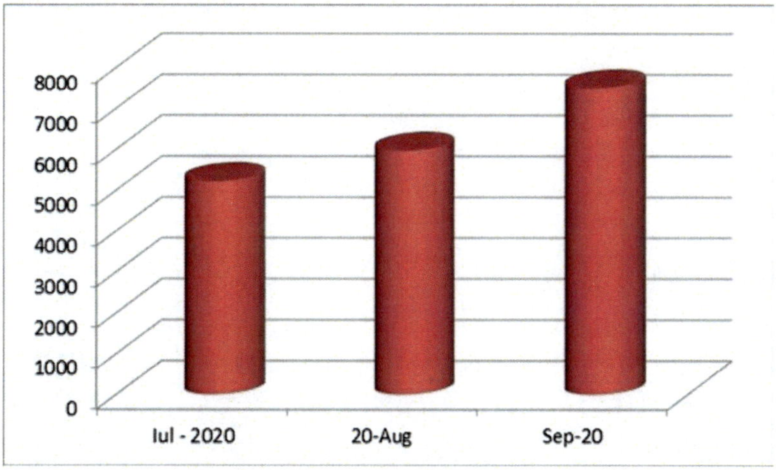

Fig. 6.7 The profit of the courier company over the 3 months analyzed. (Sources: Own calculations based on information collected from the firm)

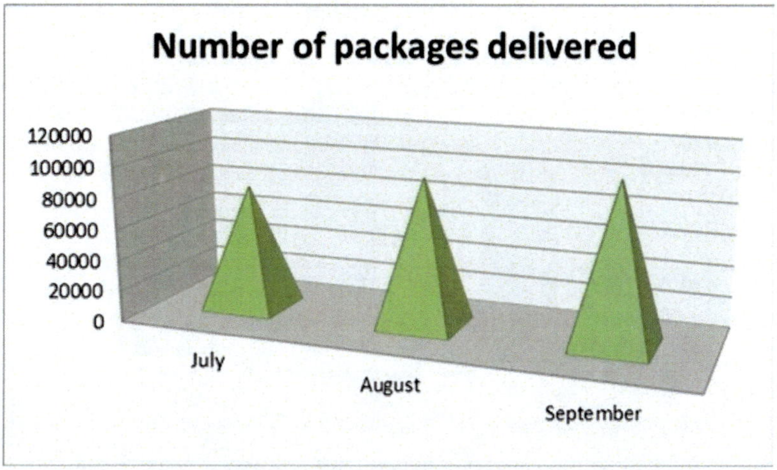

Fig. 6.8 Status of packages delivered at branch level during the period July–September 2020. (Sources: Own calculations based on information collected from the firm)

Adaptation to market changes must also be reflected in the management of society. Employee performance evaluation systems must be continuously improved (Downing, 2020). The company-wide SIAR program reflected the ability to improve the economic performance of the company. Regardless of the area of activity in which the SPC indicator is implemented, as long as it is properly adapted and used, the results will reflect an economically real situation for the company.

As regards the economic activity of the hair styling show, the economic situation is shown in accordance with the Table 6.4 below.

Table 6.4 Evolution of economic indicators for the hair-style salon (August–October 2021)

Economic indicators	Aug	Sep	Oct
Turnover	77,324	79,564	74,121
ΔQP%		2.90%	−6.84%
Profit	9924	14,574	15,279
ΔPr%		14.31	25.71
Total costs	67,400	64,990	58,842
ΔCT%		−3.58%	−9.46%
Number of employees	9	9	9
Eerfp	1.15	1.22	1.26
Eecfp	0.87	0.82	0.79
Rate of return	14.72%	22.42%	25.97%
ΔRprCT%		52.30%	15.79%
Commercial rate	12.83%	18.32%	20.61%
ΔRprQP%		42.72%	12.54%

Sources: Own calculations based on information collected from the firm

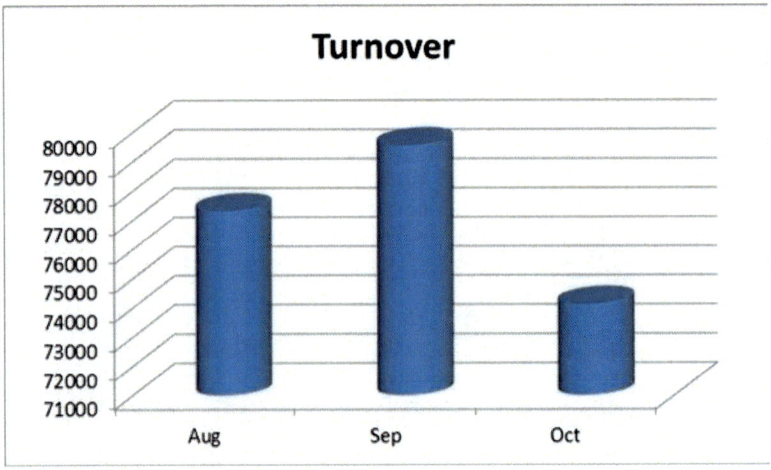

Fig. 6.9 Turnover of the hair-style salon during the period Aug–Oct 2021. (Sources: Own calculations based on information collected from the firm)

There is an increase in turnover of the firm in September 2021 compared to August 2021, an increase of almost 3%. In October, the turnover of the show fell by almost 7%, which is not a negative fact, given that the company's costs in October 2021 decreased by almost 10% compared to September of the same year (Fig. 6.9).

Concerning the main element guiding a firm's activity in a market economy, that is, its profits have been steadily increasing, so in October a profit of more than 15,000 euro was achieved, with an increase of 53% compared to August 2021. The other economic indicators used in this survey show that the company's economic activity is healthy. The profitability rate and the trade rate developed positively (Fig. 6.10).

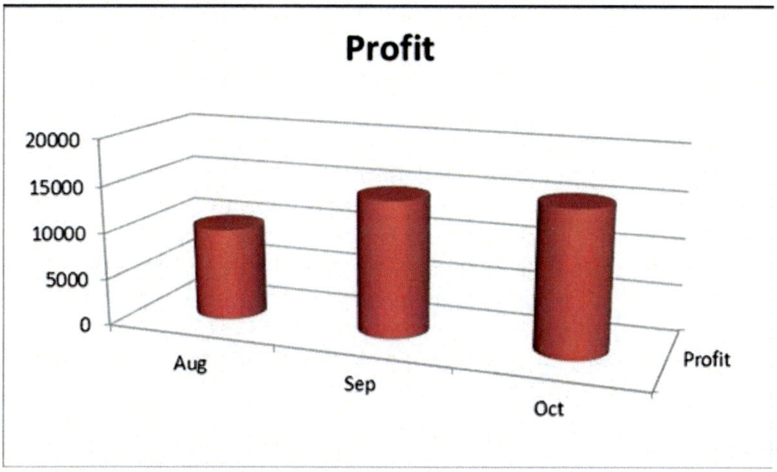

Fig. 6.10 Statement of the profit indicator in the period Aug–Oct 2021. (Sources: Own calculations based on information collected from the firm)

The elements of AR used in this program in the hair styling lounge had positive effects in terms of the business activity of the firm.

6.4 Conclusion

Even if a company manages to achieve economically positive results, it does not guarantee that it will also be successful in the market. Competition is a hallmark of the market economy, an element that sets the market in motion and continues to push economic participants toward permanent development.

The novelty elements used in this program can be applied in any sector of activity, as the study shows the introduction of these elements from the VR and AR have had a positive impact on the results of the firms analyzed.

Irrespective of the field of activity, the Spc indicator can present a real situation of the economic situation of the company vis-à-vis its clients, as long as it is properly applied. From the analysis presented in this paper, some aspects of how to implement and adapt this indicator to different companies can be distinguished. The Spc indicator analysis confirms the two assumptions underlying this work, firstly that any growth and economic development program requires continuous improvement of the economic indicators used and, secondly, in terms of performance assessment systems, innovative adjustments are needed, which of course can also use elements from other sectors of activity (Legman & Gabor, 2020b).

This study is a prerequisite for future specialist studies in this field, which will use these elements.

Finding new ways to learn and implement innovative tools is more difficult than ever today, especially when technology has taken such a momentum. This technology AR is an effective tool that can be applied in many different fields, from education to business.

At international level there are many discussions about the effectiveness of implementing such technologies in both educational programs and in the business environment. For example, at 2020 level, according to Threekit (2020), 52% of retailers think that for implementation a good partner would be needed to have a successful ROI (return on investment) after the adoption of AR. The main criteria for assessing such technologies should be the effectiveness and functionality and according to the same sources.

References

Capelle, R. (2014). *Optimizing organization design, a proven approach to enhance financial performance, customer satisfaction and employee engagement*. Jossey-Bass A Wiley Brand.

Downing W. J. (October 2020), 2019 Mid-market technology trends report. https://www2.deloitte.com/us/en/pages/deloitte-private/articles/technology-trends-middle-market-companies-survey.html

Filipowiak J. (October 2020). How can augmented reality (AR) be used for business. https://www.pixellegend.com/blog/how-can-augmented-reality-ar-be-used-for-business/

Legman, I. D., & Gabor, M. R. (2020a). *New optimization technique for sustainable manufacturing: The implementation of the Spc indicator as important element of satisfaction measurement*. Presented at the *14th international conference INTER-ENG*. Interdisciplinarity in Engineering, 8–9 October 2020. *Proceedings, 63*(1), 4, MDPI. https://doi.org/10.3390/proceedings2020063004. https://www.mdpi.com/2504-3900/63/1/4. Award – winning 1st Prize Award.

Legman, I. D., & Gabor, M. R. (2020b). Augmented reality technology – A sustainable element for industry 4.0, Acta Marisiensis. *Seria Oeconomica, 2*, 17–36.

Legman, I. D., & Gabor, M. R. *Increase economic efficiency by improving customer satisfaction*, Sapientia, Acta. Seria Philologica (in press). 15th International Conference on Economics and Business, http://csik.sapientia.ro/data/Program%20_%20Final.docx

Threekit. (2020). About Us. *Threekit Corporation*. https://www.threekit.com/about-us

Chapter 7
Monetary Policy and Current Account Balances in Nigeria (1981–2020)

Olaide Kafayat Mustapha-Jaji and Nargiza Alymkulova

7.1 Introduction

The status of a country's current account balance position has been a cause for concern to economic planners, policymakers, and academic scholars. While a surplus in the current account position is usually not a source of worry, a deficit in the current account balance has raised immense concerns, particularly in developing countries. This is because the persistent deficit in current account balances is a key indicator of low national savings and investment, the vulnerability of the domestic economy to international shocks, lack of international competitiveness, and structural economic problem such as lopsided real sector growth (Wanjau, 2014). More so, literature has identified drivers of current account balance deficit in developing countries to include deficiency in foreign exchange reserves, overrated real exchange rate, unfavorable terms of trade shocks, declining growth rate, rising domestic price level, high-yielding foreign investment, and high interest rates in developed countries (Oshota & Badejo, 2015; Osakwe & Verik, 2007). Similarly, the Nigerian economy has been characterized by excessive imports of consumables and production inputs, unstable oil prices, weak non-oil export structure, and a fragile financial system to boost foreign capital inflows as factors influencing the current account balances.

The need to address the problem of current account imbalance has solicited implementation of macroeconomic policies, particularly monetary and fiscal policy instruments. Monetary policy instruments have been noted as essential in ensuring

O. K. Mustapha-Jaji (✉)
Department of Economics, Nile University of Nigeria, Abuja, Nigeria
e-mail: 201328017@nileuniversity.edu.ng

N. Alymkulova
Nile University of Nigeria, Abuja, Nigeria

S. D. Carter (ed.), *COVID-19, Supply Chain, Climate Change, and Sustainable Development in Africa*, Springer Proceedings in Business and Economics,
https://doi.org/10.1007/978-3-031-26121-3_7

an economy's competitiveness and reducing the deficit in current account balance disequilibrium. Furthermore, literature has identified adjustment in domestic interest rate as an incentive for increasing domestic investment and capital inflows into the domestic economy, which can reduce the deficit in the current account balance.

Moreover, the Nigerian government has devalued the domestic exchange rate on different occasions and has equally adjusted the domestic interest rate and money supply. However, the extent to which these measures have influenced the current account balances has remained unresolved, due to the underrepresentation of this issue among indigenous studies in Nigeria. Thus, despite the quantum of research in both theoretical and empirical studies on issues relating to current account balances (see Iyoboyi & Muftau, 2014; Tijani, 2014; Oladipupo & Onotaniyohuwo, 2011), there exists a paucity of literature on the link between monetary policy and current account balances in Nigeria. The lack of indigenous studies on the link between monetary policy and current account balance is of immense concern due to the influence of current account balance on the foreign reserves, exchange rate, and future economic performance.

This article is divided into five sections. Section 7.1 covered the introduction, and Sect. 7.2 discussed the literature review, covering both the theoretical and empirical literature. Section 7.3 shall cover the research methods, while Sect. 7.4 shall cover the data analysis and discussion of the empirical results. Section 7.5, the final section, shall cover the conclusion and policy recommendations.

7.1.1 Research Questions

(a) What have been the developments in Nigeria's current account balance and monetary policy during the period 1981–2020?
(b) What is the impact of monetary policy on the current account balance in Nigeria for the period 1981–2020?

7.1.2 Objectives of the Study

The main objective of this study is to examine the relationship between current account balance and monetary policy in Nigeria. The specific objectives are to:

(a) Appraise current account balance positions and monetary policy development in Nigeria during the period 1981–2020.
(b) Analyze the impact of monetary policy on the current account in Nigeria over the period 1981–2020.

7.1.3 *Hypothesis Testing*

H_0: Monetary policy variables have an insignificant impact on the current account balance in Nigeria.
H_1: Monetary policy variables significantly affect the current account balance in Nigeria.

7.2 Literature Review

This involves theoretical expositions of the determinants of the balance of payment position and the review of empirical studies related to current account balances and monetary policy variables in both developed and developing countries.

7.2.1 *Theoretical Review*

The Monetary Approach The monetary approach to the balance of payment argues that disequilibrium in demand and supply of money is the principal cause of the balance of payment disequilibrium (Meade, 1951). According to this approach, a decline in money supply reduces the household's spending capacity, reducing the demand for foreign commodities. The decline in the demand for foreign commodities is expected to improve or perhaps reduce the deficit in the balance of payment. On the other hand, an increase in the money supply may deteriorate the balance of payment.

The Elasticity Approach The elasticity approach to the balance of payment stressed that distortion in the relative prices in the foreign exchange market and lack of competitiveness in international trade is a critical determinant of the balance of payment position (Rose, 1991). According to this approach, the depreciation (appreciation) of the domestic currency may enhance (weaken) exports over imports of the domestic economy, thereby influencing the position of the balance of payment.

The Balance of Payment Constraint Model The balance of payment constraint model proposed by Thirlwall (1979) argued that a deficit in the balance of payment is caused by increased national income over aggregate exports. The increase in income is expected to increase imported commodities, resulting in an unfavorable balance of payment position, which impedes economic growth. The balance of payment constraint model concluded that equilibrium or favorable balance of payment could only be achieved through an export-led growth strategy.

Absorption Approach to Balance of Payment The approach argues that balance of payment disequilibrium occurs due to imbalances between domestic absorptive and domestic productive capacity (Vines, 2008). According to this approach, a country experiences a deficit in the balance of payment position when domestic spending from economic agents (household, business firms, and government agencies) exceeds domestic production or national income. However, a country experiences balance of payment surplus when domestic production exceeds domestic spending by economic agents in the country.

7.2.2 Empirical Review

Regarding empirical studies, Uzoma and Ekine (2021) analyzed the impact of stabilization policy measures on the current account balance in Nigeria from 1980 to 2019. The study used the autoregressive distributed lag (ARDL) estimation technique, and the results showed that fiscal policy had an insignificant influence on current account balance while monetary policy variables significantly influenced current account balance in Nigeria.

Sule and Shuaibu (2020) analyzed the link between current account behavior, real exchange rate adjustment, and output growth in Nigeria. The study employed quarterly data for the period 1981Q1–2017Q4. Using the structural vector autoregression technique, the study observed that fiscal shocks significantly influenced the current account balance and output growth, while fiscal shocks had an insignificant impact on real exchange rate adjustment. Moreover, the study observed that contractionary monetary policy had a negative influence on current account balances while exchange rate shocks significantly influenced output growth. Thus, the outcome of the study supports the presence of the expenditure-switching effect. The study further observed that the influence of the shocks was found to be more noticeable during the fixed exchange rate regime than in the flexible exchange rate regime.

Efanga et al. (2020) analyzed the relationship between the balance of payment and Nigeria's economic growth from 1981 to 2010. Using the autoregressive distributed lag (ARDL) estimation method, the study observed a positive and significant relationship between the balance of payment and economic growth in Nigeria. Sani et al. (2019) examined the current account balance's impact on Nigeria's economic growth for the period 1970–2016. The study employed the autoregressive distributed lag (ARDL) estimating technique, and the results showed that the current account balance significantly enhanced economic growth in Nigeria. Moreover, they observed that the exchange rate had a negative and significant impact on the current account balance.

Longe et al. (2018) examined the impact of oil price fluctuations on current account balances in Nigeria for the period spanning 1977–2015. The study used the autoregressive distributed lag (ARDL) technique, and the findings of the study showed that in the long run, the oil price had a negative and insignificant impact on the current account balances, while in the short run, the oil price had a positive and insignificant effect on current account balances. More so, the study observed that in the short run, trade openness, gross domestic product, and population growth had an insignificant impact on current account balances, while oil price had a significant impact on current account balances in the long run.

Ogunniyi et al. (2018) investigated the impact of the current account balance on economic growth in SANE (South Africa, Algeria, Nigeria, and Egypt) countries. The study employed the autoregressive distributed lag (ARDL) and fully modified ordinary least squares (FMOLS) methods. The findings of the study showed that current account balances had a negative and significant impact on economic growth in Nigeria, Egypt, and Algeria. In contrast, the effect of current account balances on economic growth in South Africa was positive and significant.

Olanipekun and Ogunsola (2017) analyzed the impact of the exchange rate on the balance of payment (BOP) in Nigeria. Also, the study examined the impact of the exchange rate on the capital and current account balances. The study employed the autoregressive distributed lag (ARDL) method, and the findings showed that the appreciation of the exchange rate had a negative impact on the balance of payment. More so, the study found that the exchange rate had a negative impact on the current account balance while the impact of the exchange rate on the capital account was statistically insignificant.

Iyeli and Ovat (2017) analyzed the budget deficit's impact on Nigeria's current account balance for the period spanning 1970–2014. The study used the error correction modeling method and observed that budget deficit, exchange rate, and foreign private investment had a positive and significant impact on the current account balance. In contrast, the inflation rate had a negative and significant impact on the current account balance of Nigeria.

Nwanosike et al. (2017) investigated the relationship between domestic currency devaluation and balance of payment in Nigeria for 1970–2014. Using the ordinary least squares estimation technique, the study showed that domestic currency depreciation had a negative impact on the balance of payment position in Nigeria. Hjortsoe et al. (2016) examined the relationship between monetary policy and current account balance for a panel of 19 OECD countries for the period 1976 Q1–2006Q2. Using the panel estimation method, the results of the study showed that expansionary (contractionary) monetary policy would cause current account deficit (surplus) in countries with liberalized financial (product) markets.

Sadiku et al. (2015) examined the determinants of the current account deficit in FYROM. The study employed quarterly data for 1998Q1–2013Q4, and the autoregressive distributed lag (ARDL) estimation technique was utilized. The findings of the study showed that terms of trade, fiscal balance, and financial development had a positive impact on the current account balance, while trade openness had a negative impact on the current account balance.

Oshota and Adeleke (2015) examined the determinants of the current account balance in Cote d'Ivoire, Ghana, and Nigeria, for the period 1987–2008. The study employed vector autoregressive (VAR) estimation method, and the outcomes showed that domestic investment, real effective exchange rate, and relative income are significant determinants of current account balances in the long run in Cote d'Ivoire, Ghana, and Nigeria. In the short run, the study observed that domestic investment, real effective exchange rate, and relative income are insignificant determinants of the current account balance in the focused countries.

Adeyemo et al. (2015) investigated the impact of agriculture value added on the current account balance in Nigeria for the period 1980–2013. Utilizing the vector error correction modeling technique, the study observed that agricultural value added negatively impacted current account balances both in short and the long run. Moreover, the study found that gross domestic output, real effective exchange rate, and foreign assets positively influenced current account balance while per capita gross domestic product and trade terms negatively affected current account balance in Nigeria.

Oshota and Badejo (2015) analyzed the determinants of the current account balance (CAB) for a panel of West African countries for the period 1980–2012. The study used the pooled mean group (PMG) and dynamic fixed-effect (DFE) estimation techniques. The PMG estimates showed that per capita GDP, domestic investment, broad money supply, and dependency ratio are positive and significant determinants of the current account balance. At the same time, the real effective exchange rate negatively impacted the current account balance in the long run. On the other hand, the DFE estimate showed that per capita GDP and broad money supply positively and significantly affected the current account balance in the long run. In the short run, the DFE estimate showed that domestic investment had a positive impact on the current account balance, while the real effective exchange rate had a negative and significant impact on the current account balance.

Wanjau (2014) analyzed the relationship between real exchange rate, current account balance, and real income in Kenya. Specifically, the study examined the impact of the real exchange rate on current account balance and considered the Thirlwall (1979) law – whether import growth is consistent with economic growth. The study covered 1980–2011 and employed the autoregressive distributed lag (ARDL) estimation technique. The study's results showed that the exchange rate had a significant influence on current account balance, suggesting the presence of the Marshall-Lerner condition and the J-curve phenomena in Kenya. Furthermore, the study observed that Kenya's import growth is significantly greater than the level consistent with long-run economic growth.

In Turkey, Turgutlu (2014) examined the relationship between consumer loans and the current account balance over the period 2000Q1–2013Q1. The study decomposed the consumer loans into vehicle loans, real estate loans, and other loans, while macroeconomic variables such as depreciation rate, GDP growth rate, and real interest rate were included in the model. The study utilized the autoregressive distributed lag estimation method, and the results showed that real estate loans and other loans had a negative and significant impact on the current account balance. More so, the study observed that real interest rate and domestic currency depreciation had a negative and significant impact on the current account balance in Turkey.

Danmola and Olateju (2013) analyzed the relationship between monetary policy and current account components in Nigeria for 1970–2010. Using the error correction model, the study observed that monetary policy positively influenced components of the current account. Abbas et al. (2011) analyze the impact of fiscal policy on the current account balance for a panel of advanced, emerging, and low-income economies. The study employed a panel vector auto-regression estimation technique, and the results showed that fiscal policy had a positive and significant impact on the current account balance in the sampled countries. The study further noted that the relationship between fiscal policy and current account balance seems stronger in emerging and low-income economies than in advanced countries.

Udah (2010) examined the impact of macroeconomic policy, non-policy, and financial sector variables on the current account balance in Nigeria. The study used the Granger causality and the vector autoregressive (VAR) estimation techniques. The Granger causality estimate showed that the exchange rate had unidirectional causation to the current account balance, while one-way causation was observed from the current account balance to trade openness. However, the study observed no causality between measures of financial development and current account balance. The VAR estimate showed that budget deficit, monetary policy credibility, and exchange rate significantly impacted the current account balance. In addition, the study found that non-policy variables such as terms of trade, per capita income, and trade openness significantly influenced the current account balance in Nigeria.

Cevis and Camurdan (2008) analyzed the determinants of the current account balance in inflation-targeting countries for the period 1990–2006. The study employed a random effect panel estimation technique, and the study observed that economic growth, trade openness, term of trade, real exchange rate, and real interest rate differential are significant determinants of current account balances. Egwaikhide (1997) analyzed the impact of the budget deficit on the current account balance in Nigeria for the period 1973–1993. Using the ordinary least squares regression method, the study observed that budget deficit had a significant negative impact on current account balance, suggesting that budget deficit worsened the current account balance deficit in Nigeria.

7.2.3 Gaps in Literature

The literature review above clearly showed that literature abounds on the current account balance in Nigeria. These studies mainly focused on the impact of the exchange rate on current account balance (Iyoboyi & Muftau, 2014; Olanipekun & Ogunsola, 2017) and also on the impact of the current account balance on economic growth (Efanga et al., 2020; Udah, 2010). This study did not consider the link between monetary policy variables (such as money supply and interest rate) and current account balance in Nigeria. However, literature has emphasized that monetary policy variables (such as money supply and interest rate) determine current account balances and measures of correcting imbalances in a country's current account. It is against this lacuna in previous studies that this study intends to bridge the knowledge gap by carrying out a more comprehensive study on the relationship between monetary policy and current account balances in Nigeria for the period 1981–2020.

7.3 Research Methods

7.3.1 Theoretical Framework and Model Specification

This study rests on the monetary and elasticity approaches to the balance of payment position. The monetary approach emphasizes the importance of monetary policy instruments such as money supply as a key driver of the balance of payment, while the elasticity approach emphasizes the importance of exchange rate in explaining the current account balances. Thus, drawing from these approaches, this study specifies a simple model in which monetary policy variables influence current account balance. This model is given as

$$CAB = f\left(MP\right) \tag{7.1}$$

Introducing other control variables that have been identified in the literature as significant drivers of current account balances – oil price (OP) and foreign direct investment (FDI) (see (Sule & Shuaibu, 2020; Longe et al., 2018; Adeyemo et al., 2015). In this study, monetary policy is proxy by money supply (MS), domestic interest rate (DIR), and exchange rate (EXR). Incorporating these variables into Eq. (7.1), we have:

$$CAB = f\left(MS, DIR, EXR, OP, FDI\right) \tag{7.2}$$

Linearizing Eq. (7.2) and expressing Eq. (7.2) in econometric estimation form, we have:

$$CAB_t = \alpha_0 + \alpha_1 MS_t + \alpha_2 DIR_t + \alpha_3 EXR_t + \alpha_4 OP_t + \alpha_5 FDI_t + \varepsilon_t \tag{7.3}$$

To estimate Eq. (7.3), this study employs the autoregressive distributed lag (ARDL) estimation method. The ARDL helps to examine both the long-run and the short-run relationship between economic variables. More so, the ARDL technique has the advantage of combining variables which are I(1) and I(0) in the estimation process, unlike the error correction modeling and vector error correction modeling techniques which require variables to be stationary at the first difference before estimation. Thus, in line with the ARDL method, the long-run model is specified as

$$\begin{aligned} CAB_t = \beta_0 + \sum_{i=1}^{n} \beta_1 MS_{t-i} + \sum_{i-1}^{n} \beta_2 DIR_{t-i} + \sum_{i=1}^{n} \beta_3 EXR_{t-i} \\ + \sum_{i=1}^{n} \beta_4 OP_{t-i} + \sum_{i=1}^{n} \beta_5 FDI_{t-i} + \varepsilon_t \end{aligned} \tag{7.4}$$

The short-run estimation model of the relationship between monetary policy and current account balance, along with the control variables, is given below:

$$CAB_t = \delta_0 + \sum_{i=1}^{n} \delta_1 \ MS_{t-i} + \sum_{i=1}^{n} \delta_2 \ DIR_{t-i} + \sum_{i=1}^{n} \delta_3 \ EXR_{t-i} + \sum_{i=1}^{n} \delta_4 \ OP_{t-i}$$
$$+ \sum_{i=1}^{n} \delta_5 \ FDI_{t-i} + \delta_6 ecm_{t-1} + \varepsilon_t \qquad (7.5)$$

The ECM term in Eq. (7.5) is the error correction term, which measures the speed of adjustment from the short-run disequilibrium to the long-run equilibrium.

7.3.2 Theoretical/A Priori Expectation

Theoretically, it is expected that an increase in the money supply would bring about a deficit in the current account balance. An increase in the money supply would increase the consumer's purchasing power, thereby increasing imports and worsening the current account balance position. An increase in domestic interest rate (DIR) is expected to distort domestic investment, affecting exports and encouraging imports. Hence, the current account balance would deteriorate. The domestic exchange rate depreciation is expected to discourage imports and promote exports, thereby improving the current account balances. An increase in the oil price (OP) and foreign direct investment (FDI) would improve the current account position due to the inflow of foreign funds into the economy.

7.3.3 Data Measurement and Sources

Concerning the variables utilized in this study, the measurements and sources of these variables are provided in Table 7.1.

7.3.4 Estimation Techniques

Objective one is achieved using the descriptive technique, which involves graphs explaining developments and stylized facts on variables of concern. The study employed the autoregressive distributed lag (ARDL) estimation technique for objective two. The technique allows the estimation of the long- and short-run relationship between the variables under focus. In this line, the ARDL-bound co-integration estimate accommodates the estimation of variables with different orders of integration (Pesaran et al., 2001). Also, the ARDL technique automatically generates the error correction term, which measures the speed of adjustment from the short-run disequilibrium to the equilibrium in the long run.

Table 7.1 Data measurement and sources

S/N	Data (variable type)	Definition and variable measurement	Sources
1	Current account balance (CAB) (dependent variable)	The current account balance refers to the receipts (for exports) and payments (for imports) between one country and another It is measured by the ratio of the current account balance to real gross domestic investment	External sector (Section D) – CBN Statistical Bulletin, 2020 edition
2	Money supply (MS) (independent variable)	This refers to the total volume of money supplied in an economy at a given point in time. The money supply is measured by the broad money supply (M2)	Financial sector (Section A) – CBN Statistical Bulletin, 2020 edition
3	Domestic interest rate (DIR) (independent variable)	This refers to the domestic interest rate under CBN's total control. The DIR is measured by monetary policy rate, which is the benchmarks interest rate	Financial sector (Section A) – CBN Statistical Bulletin, 2020 edition
4	Exchange rate (EXR) (independent variable)	An exchange rate is the value of a country's currency against the value of other countries. The official exchange rate of Naira measures it to US Dollar (₦/US$)	External sector (Section D) – CBN Statistical Bulletin, 2020 edition
5	Oil price (OP) (independent variable)	This refers to the price of crude oil on the international oil market. The bonny light crude oil price measures oil price	CBN website http://statistics.cbn.gov.ng/cbn-onlinestats/DataBrowser.aspx
6	Foreign direct investment (FDI) (independent variable)	This refers to the inflow of foreign capital into the economy. The net inflows of FDI measure it into the Nigerian economy	External sector (Section D) – CBN Statistical Bulletin, 2020 edition

Source: Author's

Stationarity Tests To avoid spurious regression estimates, the study conducted the stationarity test, otherwise known as the unit root test, to ascertain the appropriate order of integration of the variables in the estimation model. Thus, the study utilized the augmented Dickey-Fuller (ADF) and Phillips-Perron (PP) tests.

Lag Selection Criteria To ensure appropriate selection of the lag length of both the dependent and independent variables, particularly for the short-run estimate, the lag order selection criteria (which include the FPE = final prediction error; AIC = Akaike information criterion; SC = Schwarz information criterion; and HQ = Hannan-Quinn information criterion) were employed in the study.

Diagnostics Tests The second-order test, which includes the serial correlation LM test and normality test, was conducted to guarantee a robust estimate. The insignificance of the probability values of these tests at 5% critical level implies the absence of serial correlation or autocorrelation problem in the estimated model. Also, it indicates that the estimate is appropriate for policy references.

7.3.5 Expected Contributing to Knowledge

Theoretical literature (such as the monetary approach and the elasticity approach) has stressed the importance of monetary policy in influencing the current account balance position. Thus, the findings of this study provided the stance of monetary policy as a driver of the current account balance in addition to other nonmonetary factors such as oil price and terms of trade shocks. More so, the findings of this study provided appropriate policy recommendations which will serve as a viable toolkit in managing current account balances in Nigeria.

7.4 Data Analysis and Interpretation

7.4.1 Appraisal/Stylized Facts on Current Account Balances and Monetary Policy

Appraisal/Stylized Facts on Current Account Balance Nigeria The current account balance (CAB) is an essential aspect of a country's balance of payment position, which shows the difference between a country's receipts from exports and payments for imports. Countries of the world frequently experience disequilibrium in the current account balance. However, most developing countries usually experience a deficit in the current account balance owing to excessive imports over export. Concerning Nigeria, there has been a history of deficit and surplus in the country's current account balance position. As observed in Fig. 7.1, the Nigerian economy has

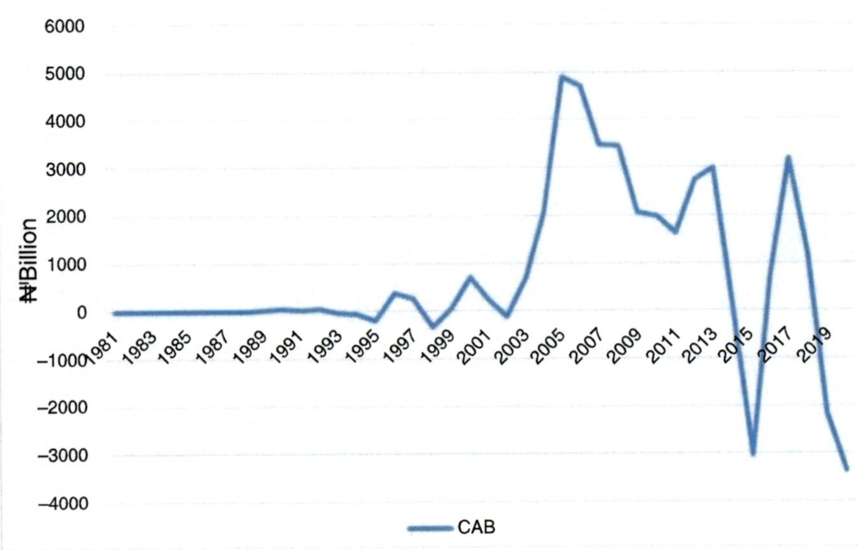

Fig. 7.1 Current account balance in Nigeria 1981–2020. (Source: Author' using data from Central Bank of Nigeria statistical bulletin, 2020 edition)

more episodes of country's account surpluses than deficits, and this is primarily due to periods of favorable oil prices in the international market. Similarly, periods of deficit in the current account balance have coincided with an era of oil price collapse and rising importation of household commodities and production inputs – petroleum products, vehicles, spare parts, clothing, drugs, and medical equipment, among others. As noted in Fig. 7.1, the Nigerian economy experienced a high surplus in the current account balances in 2005, estimated at ₦4698.05b, while the worst deficit in the current account balances was witnessed in 2020 with an estimated value of ₦-3346.51b in 2020 due to largely the impact of the current ravaging global health pandemic.

Stylized Facts on Monetary Policy Development in Nigeria Monetary policy development has majorly operated under two distinct regimes in Nigeria – the regulated regimes (pre-1986) and deregulated regimes (1986 to date). During the regulated regime, monetary policy variables such as the domestic interest and exchange rates were fixed by the monetary authority. However, during the deregulated regime, monetary policy variables are relatively determined by market forces of demand and supply. The monetary sector has experienced significant changes which influenced real variables (such as savings, investment, and gross domestic product) in the economy. Significant changes during the deregulated regimes include the reduction in maximum credit ceilings on banks in 1986, an increase in commercial bank capital base to ₦25b in 2005, and reduction in the number of commercial banks from 89 to 25 banks in 2005. Another significant adjustment was the introduction of the mon-

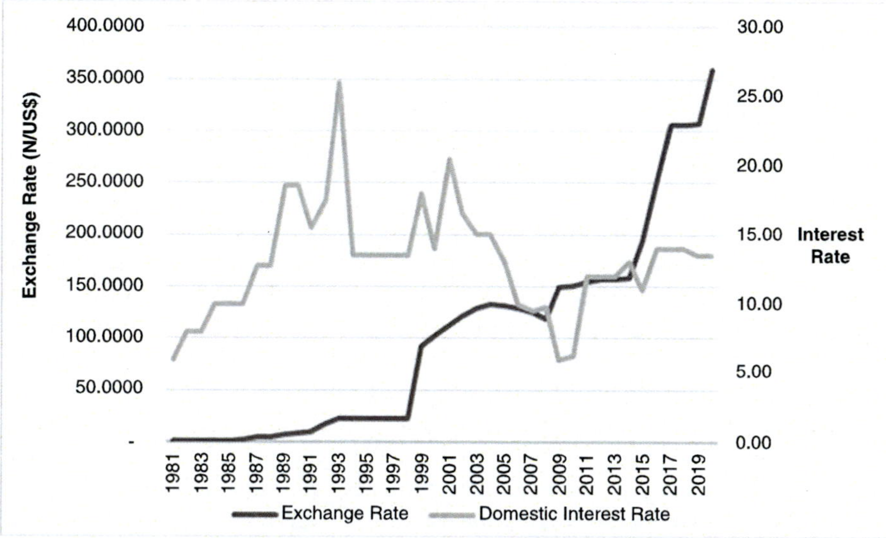

Fig. 7.2 Monetary policy variables (domestic interest rate and exchange rate) 1981–2020

etary policy rate (MPR) as a replacement for the minimum rediscounted rate (MRR), to reduce the volatility of interest rate in the money market and ultimately achieve a stable value of the domestic currency.

Despite the developments in monetary policy, the domestic interest rate (i.e., the monetary policy rate) has been relatively unstable. In contrast, the domestic exchange rate has witnessed continuous depreciation due to the official devaluation of the domestic currency against other currencies of the world. The noticeable adjustments in the monetary policy rate have usually been the decisions of the monetary policy committee (MPC) meetings of the Central Bank of Nigeria, depending on the economic realities or prevailing circumstances. Thus, the monetary policy rate currently stands at 13.50%. Similarly, continuous pressures on the exchange rate have resulted in the frequent official devaluation of the domestic currency against other foreign currencies, as witnessed in 1999, 2009, 2014, 2015, and 2020. In particular, such devaluations have resulted in the depreciation of the domestic currency against primary foreign currency, as noted in Fig. 7.2 by the upward movements of the Naira against the US Dollars, which currently stands at ₦380.50/US$1.

7.4.2 *Regression Estimate and Interpretation*

Descriptive Statistics The descriptive statistics showed that the mean values of the current account balance (CAB), money supply (MS), and domestic interest rate (DIR) are ₦661.280b, ₦7304.25b, and 13.09%, respectively, while the mean values of exchange rate (EXT), oil price (OP), and foreign direct investment (FDI) are ₦100.88, US$41.48dpb, and ₦385,760b, respectively. The median values are ₦45.54b for CAB, ₦1073.89b for MS, 13.5% for DIR, ₦107.02 for EXT, US$30.57dpb for OP, and ₦124,193b for FDI. The standard deviation showed that FDI had the highest value, while DIR had the lowest value. The skewness statistics showed that CAB is negatively skewed while the remaining variables are positively skewed. Kurtosis statistics indicated that CAB, MS, and DIR are leptokurtic, OP and FDI are platykurtic, while the exchange rate is mesokurtic, showing that the variable has a normal distribution. Lastly, the Jarque-Bera statistic rejected the null hypothesis of normal distribution for money supply (MS) and domestic interest rate (DIR). In contrast, the null hypothesis of normal distribution for the remaining variable was not rejected at the 5% critical level (Table 7.2).

Stationarity and Co-integration Tests The stationary tests, known as the unit root tests, were carried out using Phillip-Perron and augmented Dickey-Fuller tests. The results of both tests showed that all the variables were stationary at the first difference, with the exception to domestic interest rate (DIR) which was stationary at level. The mixture in the stationary status of the variables suggests the need to use the ARDL-bound co-integration method in estimating the co-integration among the variables (Table 7.3).

The results from the autoregressive distributed lag (ARDL)-bound co-integration technique showed that the value of the F-statistics is 4.97. This value is greater than the upper bound value at a 5% critical value (4.68), signifying the existence of co-integration among the variables in the estimated model (Table 7.4).

Lag Order Selection Criteria The study used the lag order selection criteria to ensure an appropriate estimation technique. It is observed from Table 7.5 that lag order one was selected as the most apt lag value for the estimated model as indicated by all the criteria. Thus, the study proceeded to estimate both the long- and short-run relationships between current account balances and monetary policy variables using the autoregressive distributed lag technique at order one.

Regression Estimate and Discussion The results from the long-run estimates showed that money supply (LMS), domestic interest rate (DIR), oil price (OP), foreign direct investment (LFDI), and exchange rate (LEXT) had the expected theoretical relationship with current account balances. More so, the study observed that money supply (LMS) and domestic interest rate (DIR) had insignificant impact on current account balances, indicating that the changes in money supply and domestic interest rate during the study period had insignificant influence on current account balances in Nigeria.

The insignificant impact of money supply on the current account balance in Nigeria can be attributed to the rising unemployment and poverty rates, which has affected the purchasing power of the households. More so, the desire to avoid inflationary pressure by the monetary authority might have moderated the supply of

Table 7.2 Descriptive statistics

Statistics/Variables	CAB	MS	DIR	EXT	OP	FDI
Mean	661.280	7304.25	13.088	100.879	41.479	385,760
Median	45.536	1073.89	13.5	107.024	30.57	124,193
Maximum	4891.74	35349.5	26	359.03	98.57	1,360,308
Minimum	−5270	14.47	6	0.61	11.28	264.3
Std. Dev.	1888.72	10788.9	3.99501	100.774	25.9291	434,063
Skewness	−0.2159	1.3744	0.6695	0.8859	0.8954	0.7718
Kurtosis	4.7559	3.5552	4.4337	2.9894	2.5685	2.1907
Jarque-Bera	5.4493	13.1065	6.4142	5.2318	5.6556	5.0631
Probability	0.0656	0.0014	0.0405	0.0731	0.0591	0.0795
Obs.	40	40	40	40	40	40

Source: Author's computation from E-views 9

Table 7.3 Stationarity test

Philips-Perron (PP) test				Augmented Dickey-Fuller test		
Variables	Level	1st Difference	Status	Level	1st Difference	Status
CAB	−1.4842	−3.4316*	I(1)	−1.2806	−6.3163*	I(1)
MS	−0.8014	−3.9436*	I(1)	−0.9460	−3.9779*	I(1)
DIR	−3.2661**	–	I(0)	−3.3098**	–	I(0)
EXR	2.3886	−4.0736*	I(1)	2.1647	−4.1245*	I(1)
OP	−1.9248	−8.2395	I(1)	−1.9248	−7.7072*	I(1)
LFDI	−2.1368	−8.3301*	I(1)	−1.8636	−8.5763*	I(1)

Source: Author's computation from E-views 9
Note: * and ** indicate 1% and 5% significant values, respectively

Table 7.4 ARDL-bound test

Test statistics	Value	
F-statistic	4.972041	
Critical value bounds	I(0)/lower bound	I(1)/upper bound
10%	2.26	3.35
5%	2.62	3.79
1%	3.41	4.68

Source: Author's computation from E-views 9

Table 7.5 Lag selection criteria

Lag	Lo		lo LR FPO	AIC	SC	HQ
0	−316.18	NA	1.4742	17.4153	17.6766	17.5074
1	−124.68	310.547*	0.00034*	9.00969*	10.8383*	9.65436*
2	−100.42	31.4668	0.00075	9.64451	13.0405	10.84176
3	−64.15	35.2929	0.00117	9.62975	14.5931	11.37957

Source: Author's computation from E-views 9. * indicates lag order selected at 5% significance level
FPE final prediction error, *AIC* Akaike information criterion, *SC* Schwarz information criterion, *HQ* Hannan-Quinn information criterion

money in the economy. The insignificant impact of domestic interest rate can be attributed to the high interest rate, which has reduced the level of investment in the country. Also, the economic and political instability may have contributed to reducing the potency of the interest rate policy in influencing the current account balance position in Nigeria. The insignificant impact of money supply on the current account balance contrasts with the findings of Oshota and Badejo (2015) and Danmola and Olateju (2013).

On the other hand, oil price (OP) and foreign direct investment (LFDI) and exchange rate (LEXT) had a positive and significant impact on current account balances in Nigeria. The significant influence of oil price on current account balance

Table 7.6 Regression estimate

Regressors	Estimated co-efficient	Standard error	t-Statistics	Prob.
Long-run regression estimate				
C	−0.01065	0.06278	−0.16961	0.8666
LMS	−0.01738	0.01015	−1.71291	0.0986
DIR	−0.01227	0.01495	−0.82036	0.4195
OP	0.00072	0.00032	2.26697**	0.0319
LFDI	0.01634	0.00438	3.73158*	0.0007
LEXT	0.01891	0.00628	3.00921*	0.0048
Short-run regression estimate				
ECM(−1)	−0.81247	0.19887	−4.08542*	0.0004
Δ(CAB(−1))	0.46434	0.17794	2.60941*	0.0148
Δ(LMS(−1))	−0.05053	0.03887	−1.30004	0.2050
Δ(DIR)	−0.00316	0.00156	−2.03340	0.0523
Δ(OP)	0.00058	0.00029	2.03751	0.0519
Δ(LFDI)	0.02219	0.00903	2.45801**	0.0210
Δ(LEXT)	0.00997	0.01162	0.85711	0.3992
$R^2 = 0.7690$ Adjusted $R^2 = 0.6714$			F-stat. (Prob.) = 7.87 ($p < 0.05$) Durbin-Watson = 1.7289	

Source: Authors' computation from E-views 9, 2020. * and ** signifies 1% and 5% level of significance

can be attributed to the importance of oil revenue as the primary source of foreign exchange earnings in Nigeria, while the significance of the exchange rate can be attributed to the rising depreciation of the domestic currency resulting from the escalating importation of both households and manufacturing inputs in Nigeria. Finally, the significance of foreign direct investment may be attributed to the rising trend in the inflow of foreign capital into the Nigerian economy over the years. The outcome of a significant and positive impact of the exchange rate on current account balances is in line with Iyeli and Ovat (2017) but in contrast to Olanipekun and Ogunsola (2017) and Nwanosike et al. (2017).

With respect to the short-run estimate, it was observed from Table 7.6 that the first lagged value of current account balances (ΔCAB(−1)) and the current value of foreign direct investment (ΔLFDI) had a positive and significant impact on current account balances, while the first lagged value of money supply (ΔLMS(−1)) and current values of domestic interest rate (ΔDIR), oil price (ΔOP), and exchange rate (ΔLEXT) had an insignificant impact on current account balances in the short run. Furthermore, the short-run estimate indicated that the error correction term had the expected negative signs. The error correction term with the coefficient of −0.812 connotes that the model adjusts its short-run disequilibrium by 81.2% speed of adjustment toward the long-run equilibrium.

Furthermore, the coefficient of multiple determinations (R^2) showed that the model had a relatively good fit with about 76.9% of variations in current account balances being explained by the explanatory variables while the remaining 23.1% of variations in current account balances are explained by variables outside the

Table 7.7 Diagnostic tests

Test	F-Statistics	Obs*R-squared	Prob. F(2, 24)	Prob. chi-square(2)
Serial correlation LM	0.43753	1.33678	0.6507	0.5125

Source: Author's computation from E-views 9, 2020

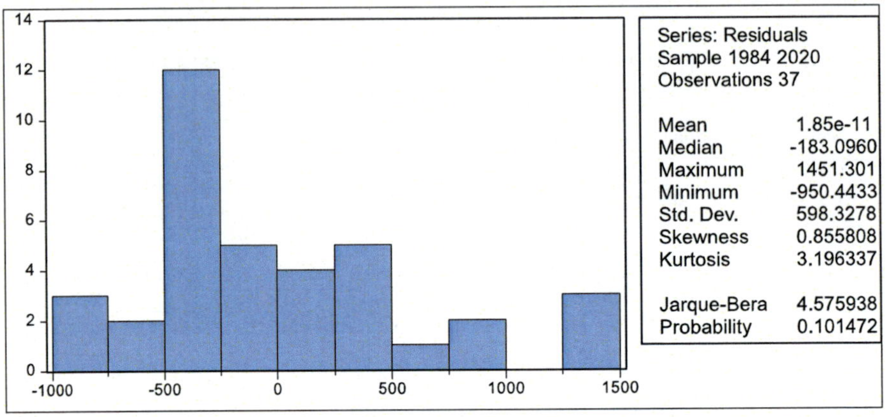

Fig. 7.3 Normality test. (Source: Author's computation from E-views 9, 2020)

model. The Durbin-Watson statistics of 1.729 showed that the regression estimate is relatively free from autocorrelation problem, thereby suggesting that the estimate is suitable for policy references. The result of the Durbin-Watson statistics is also supported by the serial correlation LM and normality estimates in Table 7.7 and Fig. 7.3, respectively. Both estimates were insignificant at 5%.

7.5 Conclusion and Recommendation

This study examined the relationship between monetary policy and current account balance in Nigeria for the period spanning 1981–2020. Employing the autoregressive distributed lag model, the study observed that monetary policy proxied such as money supply and domestic interest rate had insignificant impact on current account balance while monetary policy variable proxied by exchange rate had a significant impact on the current account balance. Thus, the study concluded that the impact of monetary policy on current account balance depends on the measures of monetary policy employed. Consequently, policy must be designed to strengthen macroeconomic shock absorbers in the country.

Demand management policies should be encouraged to enhance the performance of the real sector of the economy. Finally, the monetary authority should put in place investment plans to facilitate capital inflows, such as a 15% interest rate on foreign deposits in the domestic financial institution with a duration of 5 years and a 20%

interest rate on deposits between 5- and 20-year duration. These measures would contribute to the inflow of more foreign exchange into the economy and reduce the negative pressure on Nigeria's current account balance position.

References

Abbas, S. M. A., Bouhga-Hagbe, J., Fatas, A., Mauro, P., & Velloso, R. C. (2011). Fiscal policy and current account. *International Monetary Financial (IMF) Economic Review, 59*(4), 603–629.

Adeyemo, T. A., Ajiola, S., Odetola, S. K., & Okoruwa, V. O. (2015). Impact of agricultural value added on current account balances in Nigeria. *Journal of Economics and Sustainable Development, 6*(4), 108–117.

Cevis, I., & Camurdan, B. (2008). The determinants of the current account balance in inflation targeting countries. Available at https://www.researchgate.net/publication/23533393

Danmola, R. A., & Olateju, A. O. (2013). Impact of monetary policy on the current account balance in Nigeria. *ISOR Journal of Humanities and Social Science (IOSR-JHSS), 7*(3), 67–72.

Efanga, U. O., Etim, R. S., & Jeremiah, M. S. (2020). The impact of balance of payment on economic growth in Nigeria. *Journal of Applied Financial Economics, 1*(2), 145–160.

Egwaikhide, F. O. (1997). Effects of budget deficit on the current account balance in Nigeria: A simulation exercise. *African Economic Research Consortium (AERC)* Research paper seventy (70).

Hjortsoe, I., Weale, M., & Wieladke, T. (2016). Monetary policy and current account: Theory and evidence. Discussion Paper, *External MPC Unit*, Bank of England, No. 45.

Iyeli, I. I., & Ovat, O. O. (2017). Budget deficits and current account disequilibrium in Nigeria: A econometric investigation. *International Journal of Development and Economic Sustainability, 5*(4), 31–43.

Iyoboyi, M., & Muftau, O. (2014). Impact of exchange rate depreciation on the balance of payments: Empirical evidence from Nigeria. *Cogent Economics & Finance, 2*(9), 1–23.

Longe, A. E., Adelokun, O. O., & Omitogun, O. (2018). The current account and oil price fluctuation nexus in Nigeria. *Journal of Competitiveness, 10*(2), 118–131.

Meade, J. (1951). *The balance of payments*. Oxford University Press.

Nwanosike, D. U., Uzoechina, B., Ebenyi, G. O., & Ishiwu, V. (2017). Analysis of balance of payment trend in Nigeria: A test of Marshall-Lerner hypothesis. *Saudi Journal of Business and Management Studies, 2*(5A), 468–474.

Ogunniyi, M. B., Iwegbu, O., & Adekoya, K. I. (2018). A comparative analysis of the impact of current account balances on economic growth of SANE countries. *International Journal of Development and Sustainability, 7*(3), 1220–1237.

Oladipupo, A. O., & Onotaniyohuwo, F. O. (2011). Impact of exchange rate on balance of payment in Nigeria. *Africa Research Review: An International Multidisciplinary Journal, 5*(4), 73–88.

Olanipekun, D. B., & Ogunsola, A. J. (2017). Balance of payment crisis in Nigeria: The role of exchange rate. *International Journal of Economics, Commerce and Management, 5*(5), 119–130.

Osakwe, P, & Verik, S. (2007). Current account deficit in Sub-Saharan Africa. Do they matter? United Nations Commission for Africa, 201–220.

Oshota, S. O., & Adeleke, I. A. (2015). Determinants of the current account balance in Nigeria, Ghana and cote d'Ivoire. *Acta Universitatis Danubius Oeconomica, 11*(3), 127–145.

Oshota, S. O., & Badejo, A. A. (2015). What drives current account balance in West Africa states: Evidence from panel ARDL. *Journal of International and Global Economic Studies, 8*(2), 91–105.

Pesaran, M. H., Shin, Y., & Smith, R. J. (2001). Bounds testing approaches to the analysis of long-run relationships. *Journal of Applied Econometrics, 16*, 289–326.

Rose, A. K. (1991). The role of exchange rates in a popular model of international trade: Does the 'Marshall-Lerner' condition hold? *Journal of International Economics, 30*(3–4), 301–316. https://doi.org/10.1016/0022-1996(91)90024-Z

Sadiku, L., Fetahi-Vehapi, M., Sadiku, M., & Berisha, N. (2015). The persistence and determinants of current account deficit of FYROM: An empirical analysis. 7th international conference, the economies of Balkan and Eastern Europe countries in the changed world, EBEEC 2015, May 8–10, 2015. *Procedia Economics and Finance, 33*, 90–102.

Sani, G. K., Musa, A. U., & Sani, Z. (2019). Current account balance and economic growth in Nigeria: An empirical investigation. *Central Bank of Nigeria (CBN) Journal of Economic and Financial Review, 57*(2), 83–106.

Sule, I. K., & Shuaibu, M. (2020). Current account behavior, real exchange adjustment and relative output in Nigeria. *Journal of Economic Development, 45*(3), 77–99.

Thirlwall, A. (1979). The balance of payment constraint as an explanation of international growth rate differences. *BNL Quarterly Review, 32*(128), 45–53.

Tijani, J. O. (2014). Empirical analysis of balance of payment adjustment mechanisms: Monetary channel in Nigeria, 1970–2010. *Mediterranean Journal of Social Sciences, 5*(4), 67–76.

Turgutlu, E. (2014). Current account balance implications of consumer loans: The case of Turkey. *International Journal of Applied Economics, 11*(2), 55–64.

Udah, E. B. (2010). Adjustment policies and current account behavior: Empirical evidence from Nigeria. *Global Journal of Social Sciences, 9*(1), 45–62.

Uzoma, C. N., & Ekine, D. I. (2021). Stabilization policy measures and current account balance in Nigeria: An autoregressive distributed lag (ARDL) approach. *IIARD International Journal of Economics and Business Management, 7*(2), 17–32.

Vines, D. (2008). Absorption approach to the balance of payment. In S. N. Durlauf & L. E. Blume (Eds.), *The new Palgrave dictionary of economics*. Palgrave Macmillan. https://doi.org/10.1007/978-1-349-58802-2_3

Wanjau, B. M. (2014). The relationship among real exchange rate, current account balance and real income in Kenya. *International Journal of Business and Social Science, 5*(9), 97–118.

Part IV
Entrepreneurship

Chapter 8
Insights into African-Led Food Start-Ups and How They Are Thriving Using a Qualitative Approach

Pearl Ogekuwor Kitcher, Sylvester Asiedu, and Josephine Yeboah

8.1 Background

The African population is about 60% youthful, and there is chronic unemployment among the youth on the continent (Awogbenle & Iwuamadi, 2010; Jardim, 2021). On average, a graduate is likely to land their first paying job 3 years after graduation (Arthur-Holmes et al., 2022). Therefore, most graduates resort to taking more academic courses or starting a business to survive harsh economic situations on the continent.

Due to the problem of chronic youth unemployment on the continent, nearly one in every three youthful Africans has either initiated a business or started a side hustle (Kanu, 2020). Despite the springing up of multiple companies in Africa, African entrepreneurs encounter many challenges. Due to these challenges, most businesses shut down within the first year of operation (Adu et al., 2020). This survey was conducted to identify the obstacles that African entrepreneurs face. A Deep Dive series was initiated on YouTube to get to know the experiences of African entrepreneurs, especially people of Ghanaian origin, and identify the barriers to scaling up their businesses.

P. O. Kitcher (✉)
Chocolart Haus, Greater-Accra, Ghana; MSc. Public Health Nutrition graduate, University of Chester, Chester, UK

S. Asiedu
Information Systems Consultant and Startup Entrepreneur, Alumnus of the University of Ghana Business School, Greater Accra, Ghana

J. Yeboah
University of Chester, Chester, UK

S. D. Carter (ed.), *COVID-19, Supply Chain, Climate Change, and Sustainable Development in Africa*, Springer Proceedings in Business and Economics,
https://doi.org/10.1007/978-3-031-26121-3_8

8.2 Ghana's Cocoa Sector

More than 60% of the world's cocoa is supplied by Ghana and Cote d'Ivoire (Boysen et al., 2021); however, this cash crop is vastly underutilized in these countries (Van Huellen & Abubakar, 2021). Cocoa from these countries is primarily exported as raw or semi-processed as cocoa butter, cocoa powder, or cocoa liquor. Only small quantities of cocoa are processed locally from cocoa beans into chocolate bars. Aside from making chocolate, other by-products of the cocoa value chain are also not utilized to make valuable products (Hütz-Adams et al., 2016). Other components of cocoa, including the pod, the pulp, and the husk, are all minimally utilized; meanwhile, they can be processed into products of economic value.

8.3 The Confectionery Space in Ghana

The cake sector witnessed a high demand during the Covid-19 pandemic season. This time was accompanied by an unprecedented lockdown that lasted for an extended period; people lost their jobs, and loved ones died from the disease (Posel et al., 2021). People found the need to appreciate their relationships and show love to people that matter to them (Ahuja & Khurana, 2021). In light of this, orders for cakes escalated, and young people in Africa leveraged this demand for cakes to learn baking and make customized orders for all occasions.

8.4 Food Staples, Food Quality, and Agricultural Technologies in Ghana

Over the years, agricultural techniques have transitioned, and currently, farmers use strong chemicals to produce food crops, get rid of pests and weeds, and increase crop yields. These have led to the introduction of modern high-end technologies so that harvesting can be done much more easily. One such technology that is helping to create quality foodstuffs, especially vegetables, is using hydroponics (Samarakoon et al., 2019).

8.4.1 Aims

This study aimed at investigating and getting insights from African-owned start-ups, the motivation behind them, their business model, and how they are thriving, either locally or internationally. The insights were shared with people who have similar interests or would want to venture into entrepreneurship in Africa and its environment.

8.4.2 *Methodology*

Start-up founders who participated in this study were primarily young, between 18 and 35 years at the time the data was gathered. Participants of this cross-sectional study were of African descent, precisely Ghana, and all had an undergraduate degree as the minimum qualification. Some interviews can be viewed on the YouTube channel bearing the username Sylvester Asiedu.

8.4.3 *Findings*

Some of these themes were identified based on organized virtual one-on-one interviews. From interacting with the business owners, their motivation for beginning was to solve a problem. These problems presented themselves as opportunities. Start-ups in Africa mostly get the raw materials locally, but the packaging and other processing equipment are sourced internationally. Funding is a challenge that many African-led entrepreneurs face. Some barriers that hinder scalability for African-led start-ups include difficulties in mass production and getting the needed human resource to fill significant skills gaps in their businesses (Table 8.1).

8.5 Selected Case Studies

8.5.1 *Chocolart Haus*

Chocolart Haus is an artisanal chocolate business that adds aesthetic value to indigenous Ghanaian chocolate. Chocolart Haus aims to create globally competitive brands of artisanal chocolate and, through this, will also create livelihoods for people in Ghana. The company makes handmade chocolate bars, assorted chocolate truffles, chocolate-coated cookies, and chocolate fountains for special occasions (Fig. 8.1).

8.5.2 *Wrapped Gh*

Wrapped Gh. is a small business that began providing wrapping of custom-made gifts for people during the Covid-19 pandemic. The products are edible and nonedible, some of which are made locally and some of which are imported into the country (Fig. 8.2).

Table 8.1 Themes identified by entrepreneurs

Themes	Answers
Motivation to start the business	Passion for having access to high-quality fruits and vegetables. The vision to add value to locally made Ghanaian chocolate. I decided to start my business because I would stay a year at home while applying to medical school for the next academic year. I started it out of boredom, although I had been interested in entrepreneurship for a while.
Challenges along the way	Cybersecurity issues such as the IG page being hacked and restoring it costs money. Capability of boosting consumer confidence in the quality of fruits and vegetables in the current market. Challenges with setting up a strong working team of professionals.
Current business challenges	Sourcing manufacturing tools and packaging materials locally. Achieving schoolwork and business balance. Access to finance to scale up production.
Vision for the next 5 years	Access funding to be able to increase production capacity and efficiency. Scaling up from retail to wholesale.
SDG that the business is tackling	SDG 2 zero hunger. SDG 8 decent work and economic growth.
Words to a person considering starting a business in Africa	There are so many untapped opportunities that a yet to be explored. I will advise them to put much effort into effective research in whichever field they choose. Have a backup plan.

8.6 Tropical Growers

Some entrepreneurs started a fruits and vegetables growing company to address SDG 2 Zero Hunger. The business uses modern methods and greenhouses, as shown below. They also implemented the use of hydroponic technologies to grow fresh vegetables, which are periodically supplied to selected retail shops in Ghana (Fig. 8.3).

8.7 Conclusion

There is a need to develop solutions to help African-led businesses to scale up to curb the problem of youth unemployment and grow African macroeconomies. Solutions that aim to equip entrepreneurs in emerging economies with the needed business support will help them build resilience, scale up, and become sustainable.

There are significant barriers to access to capital on the African continent. Even though there are business incubators that offer business growth support through training, there is still a considerable challenge with accessing funds for African

Fig. 8.1 Examples of chocolate products made by Chocolart Haus. More details can be found about Chocolart Haus on Instagram @chocolarthaus, on Youtube @ChocolArt, and on Facebook @_chocolart

businesses. Bank loans have high interest rates and ridiculous checklists of eligibility requirements, making it difficult for start-ups to access these opportunities.

There are also some scaling-up challenges. For instance, there are difficulties in obtaining job-fit of human resources, obtaining equipment, and obtaining raw materials, which are imported at a high rate.

References

Adu, I. N., Boakye, K. O., Suleman, A. R., & Bingab, B. B. B. (2020). Exploring the factors that mediate the relationship between entrepreneurial education and entrepreneurial intentions among undergraduate students in Ghana. *Asia Pacific Journal of Innovation and Entrepreneurship, 14*, *215*.

Ahuja, K. K., & Khurana, D. (2021). Locked-down love: A study of intimate relationships before and after the COVID lockdown. *Family Relations, 70*(5), 1343–1357.

Fig. 8.2 Examples of wrapped products and gifts from Wrapped Ghana. More details of Wrapped Gh. can be found on Instagram @wrapped_gh

Fig. 8.3 Examples of farm produce and farming methods implemented by tropical growers. More details about tropical growers can be found on Instagram @tropicalgrowers

Arthur-Holmes, F., Busia, K. A., Vazquez-Brust, D. A., & Yakovleva, N. (2022). Graduate unemployment, artisanal and small-scale mining, and rural transformation in Ghana: What does the 'educated youth involvement offer? *Journal of Rural Studies, 95*, 125–139.

Awogbenle, A. C., & Iwuamadi, K. C. (2010). Youth unemployment: Entrepreneurship development programme as an intervention mechanism. *African Journal of Business Management, 4*(6), 831.

Boysen, O., Ferrari, E., Nechifor, N., & Tillie, P. (2021). *Impacts of the cocoa living income differential policy in Ghana and Côte d'Ivoire (no. JRC125754)*. Joint Research Centre (Seville site).

Hütz-Adams, F., Huber, C., Knoke, I., Morazán, P., & Mürlebach, M. (2016). *Strengthening the competitiveness of cocoa production and improving the income of cocoa producers in West and Central Africa*. SÜDWIND eV Kaiserstr.

Jardim, J. (2021). Entrepreneurial skills to be successful in the global and digital world: Proposal for a frame of reference for entrepreneurial education. *Education Sciences, 11*(7), 356.

Kanu, C. C. (2020). The context of Igwebuike: What entrepreneurship development systems in Africa can learn from the Igbo apprenticeship system. *AMAMIHE Journal of Applied Philosophy, 18*(1), 2019.

Posel, D., Oyenubi, A., & Kollamparambil, U. (2021). Job loss and mental health during the COVID-19 lockdown: Evidence from South Africa. *PLoS One, 16*(3), e0249352.

Samarakoon, U., Elder, D. R., Crook, N. C., Shaffer, J., & Lieb, M. (2019). Interdisciplinary project-based learning: Developing low-cost hydroponic Systems for Subsistence Farmers in Ghana. *NACTA Journal, 63*(1A), 1–6.

Van Huellen, S., & Abubakar, F. M. (2021). Potential for upgrading in financialised Agri-food chains: The case of Ghanaian cocoa. *The European Journal of Development Research, 33*(2), 227–252.

Chapter 9
Women Entrepreneurs in African Countries

Andrea Smith-Hunter, Daniel Robeson, and Gabrielle Hunter

9.1 Introduction

Across the globe, businesses owned and operated by women have been said to constitute close to twenty-five to thrity-three percent, of all businesses owned worldwide (Mwatsika, 2015). It does not stop here; these businesses owned by women are also said to be growing at a rate faster than any other businesses, as well as the overall economy in any one country. Entrepreneurship, by women, regardless of the corresponding attribute – industry, socioeconomic status, race, ethnicity, or type or size of business – has played and continues to play an essential role in the development of many countries. These essential roles are also said to have led to the continued development and future transformation of countries they have touched, including those countries on the African continent. A good place to ultra-focus on women entrepreneurs is to study this monumental and far-reaching area from a variety of these African countries.

Many general studies have been done on women entrepreneurs from Africa. At first, these often looked at an assessment of the marketing practices among African and Caribbean small businesses in London (Stokes et al., 2015). The findings reveal the patchy application of an adapted market orientation framework; they also indicated "informal" marketing deliberations taking place and is consistent with Stokes and Blackburn's (Stokes et al., 2015) assertion that marketing in the small business

A. Smith-Hunter (✉)
Siena College, New York, NY, USA
e-mail: ahunter@siena.edu

D. Robeson
Siena College, New York, USA

G. Hunter
University of Toronto, Toronto, ON, Canada

S. D. Carter (ed.), *COVID-19, Supply Chain, Climate Change, and Sustainable Development in Africa*, Springer Proceedings in Business and Economics,
https://doi.org/10.1007/978-3-031-26121-3_9

sector seems to be an informal and unplanned activity that relies on the intuition and energy of the owner-manager. This combination of both official and unofficial divisions of business are termed "dual economies" (Spring, 2009: 20). This concept implies business in Africa can be appraoached with a binary perspective, but this fails to acknowledge the educational and various kinds of capital African entrepreneur in the economic sphere (Spring, 2009). Moreover, restricted access to capital, business connections, commerce niches, and commodity management prevented upward trajectories for some businesses, so this lead to the endorsement of entrepreneurship—especially for women in business. The new generation of African entrepreneurs forms an endpoint of the continuum because their global business methods, networks, financial transparency, and business ethics propel them to success (Spring, 2009).

Overall, studies on women entrepreneurs in Africa looked at how and why gender issues are important when considering strategies to improve Africa's competitiveness in the world (Bardasi et al., 2006). Women entrepreneurs from this geographic area are more likely than their male counterparts to be engaged in family enterprises (Bardasi et al., 2006). The same study found that the women entrepreneurs are generally younger and less likely to be married (Bardasi et al., 2006). A similar study that also looked generally at women entrepreneurs in Africa cited the main reasons and difficulties to establish businesses and identified possible differences between those that go into industry and those that establish commerce or service businesses (Machado et al., 2016). Some main reasons and difficulties identified for women to establish businesses were lack of financial capital, lack of human capital, and lack of network capital (Machado et al., 2016). Identifying possible differences between those that go into industry versus those that establish commerce or service businesses were shown as having lesser versus greater of these types of capital (Machado et al., 2016; Brixiova et al., 2018). Another study examined women entrepreneurs in different African countries overall, and found that a determining factor in the success of their subsequent businesses is how they combined these three different types of capital into a matrix of success (Markovi, 2007).

The main objectives of this chapter operate along the lines of the above premise and advocate that women's entrepreneurship will continue to play a major role in the development and economic growth of many African nations' economies. It purports that the place of women entrepreneurs in any economic sphere is a role that is central and that the cyclical effect of their operations in an economy is often far-reaching and long-lasting. With that in mind, there is a perpetuation of the reality that the very nature of women's entrepreneurship presents itself as a catalyst for the instigation and development of an important component of a country's gross domestic product and ultimately as a source of income and survival for families. What then does the chapter do in its contribution to the realm of entrepreneurship in general and women entrepreneurship in particular? It outlines objectives that focus on key literature that has been developed to study the genre and emphasizes the barriers, challenges, and motivations of the said women entrepreneurs. This lineage along gender lines, entrepreneurship lines, and economic development lines presents a critical confrontation at a time when the world of work is evolving and income attainment is morphing into dimensions beyond the ordinary. As such, it offers the unique opportunity to display the critical role that women as entrepreneurs can contribute to their households, communities, and countries' economies.

9.2 Literature Review on Women Entrepreneurs in Various African Countries

The sections below offer a detailed look at the women entrepreneurs in various African countries. It highlights the challenges, motivations, and problems faced by women entrepreneurs. It emphasizes the potential for economic growth and the legitimate place entrepreneurship offers as an avenue for economic growth for women on this continent.

9.3 Challenge, Obstacles, and Problems

Panda (2018) did a study that identified and ranked constraints faced by women entrepreneurs in developing countries such as Botswana, said to arise from gender discrimination, work-family conflict, as well as political, infrastructural, educational, and economic challenges (Panda, 2018).

Next to women entrepreneurs in South Africa, sub-Sahara, and Sudan, where they encountered a myriad of issues and problems as they operated their businesses. One study looked at business owners' action planning because the functionality of planning is controversial in entrepreneurship research and there is relatively little psychological research on planning in these countries (Frese et al., 2007).

While what determined the types of problems encountered by small businesses in Nigeria and why so many failed, despite the programs established by the government and nongovernmental organizations, is often explored, a definitive link is sometimes elusive. A positive relationship has been documented between small-business development and economic growth in developed countries (Okpara et al., 2007). Such important topics are also explored by Northern Sudanese female entrepreneurs (Welsh et al., 2013). Among many important topics, they researched what were the major challenges North Sudanese female entrepreneurs faced in (a) venture startups and (b) running their businesses. Personal issues, management skills, and obtaining financing are challenges, while sources of support including schooling and previous work experience, moral support, and institutional support were identified as major areas the women entrepreneurs lacked (Welsh et al., 2013).

Obstacles and problems faced by women entrepreneurs in Tanzania, Tunisia, and the UAE are well documented in several vital areas. Lindvert et al. (2015) investigated how women entrepreneurs in Tanzania assess their accessibility to different external financial sources and what they prefer as their financial source (Lindvert et al., 2015). A comparison was made between MENA countries and women entrepreneurs in Saudi Arabia and looked at how easy it is to do business for women in those countries compared to Saudi Arabia (Minkus-Mckenna, 2009). The authors found that the ease of doing business was difficult, almost impossible for women entrepreneurs (Minkus-Mckenna, 2009). In contrast, Guérin (2006) explored women's financial practices and the causes that drove them to want to be business owners. This study suggests that Senegalese women entrepreneurs were interested in attaining financial self-independence and finding opportunities to earn income (Guérin, 2006).

Uganda, women entrepreneurs faced significant challenges whilst starting their own businesses (Stevenson, 2005), systematic research on the precise nature of these barriers for Ugandan women is somewhat limited. However, sufficient evidence indicated they have difficulty accessing financing due to collateral constraints; they have unequal access to land and property titles through matrimonial and inheritance laws; they are challenged by patriarchal legislation, which limit their ability to borrow money as well as their independence. These infringements on their rights, coupled by financial and market illiteracy, hinder their entrepreneurial opportunities, further exacerbated by the limiting beliefs on women culturally (Stevenson, 2005). The study also found that lack of education limits the scope of women's enterprise activity and is one of the largest causes of poverty for Ugandan households (Stevenson, 2005). Women with higher education have more capacity to develop growth enterprises and lead the route out of poverty (Stevenson, 2005).

In Zimbabwe, the success and struggles faced by the women entrepreneurs were laboriously documented in a study by Mboko and Smith-Hunter (2009). The authors found that female business owners generally lacked business training and experience but still went on to establish businesses, but this did not stop them from entering the business world. Motivation for the women entrepreneurs came from the need to support their families (Mboko and Smith-Hunter, 2009; Nani, 2011).

Research on why small-business ventures in Nigeria are short-lived and not sustainable was completed by Adisa et al (2014). Four factors were found to be the major challenges confronting small businesses in Nigeria: lack of adequate funding, poor record-keeping and information management, inability to distinguish business capital from personal money, lack of crucial infrastructural facilities, and lack of proper business and management skills/knowledge (Adisa et al, 2014). While Tengeh et al. (2017) analyzed a group of women entrepreneurs in Kigali, Rwanda to find specific barriers they faced were the following: a lack of collateral to obtain loans, high taxes, a lack of information technology skills and access, high-interest rates, high transport costs, and a lack of entrepreneurial skills, but some barriers are also compounded by cultural and psychological factors, which were all seen as long-standing barriers (Tengeh, et al., 2017).

Richardson et al. (2004) looked at reasons why female entrepreneurs struggle to create businesses in Africa. Some constraints to growth that were said to significantly do so are finance, intense competition, and dealing with corruption among regulatory officials (Richardson et al., 2004), while a study by Abdiraham (2016) observed the challenges and motivations of 180 women entrepreneurs in the

Somali Region of Ethiopia. The women entrepreneurs needed demand creation in the area of microfinance loan applications. The women entrepreneurs, for the most part, started their businesses with their savings and loans from family and friends, but such limited constraints served as a deterrent to the continuation of their businesses (Abdiraham, 2016). The authors concluded that women entrepreneurs need demand creation in the area of microfinance loan applications (Abdiraham, 2016).

Goedhuys and Sleuwaegen (2000) who looked at women entrepreneurs in Côte d'Ivoire analyzed the factors that make inhabitants determine if they want to enter entrepreneurship and self-employment. Financial constraints were found to play a role in restraining entrepreneurship growth and firm growth. This article looked at factors that played a critical role in entrepreneurial success in Cote D'Ivoire. Out of many factors, entrepreneurial activity was suggested to play a key role in positive outcomes. The learning process was said to take place both before and after entry into the industry as firms grow to larger sizes (Goedhuys & Sleuwaegen, 2000). Fofana et al. (2015) who also studied the effect of microcredit on women's livelihood and empowerment in rural areas of Côte d'Ivoire did a cross-sectional survey, which was conducted among 185 borrowers from 2 microfinance institutions and 209 non-borrowers in rural areas of Côte d'Ivoire (Fofana et al., 2015). A post-survey group discussion was conducted between both subject groups. After matching the two groups on the probability of obtaining credit, it was found that microcredit groups had on average a higher income and a higher value of household assets than the non-borrowers (Fofana et al., 2015).

Studies from Strobbe et al. (2015), Singh et al. (2008), Hailemariam (2018), Belwal et al. (2011), Mulatu et al. (2018), Hundera et al. (2019), and Wube (2010) provide a comprehensive picture of women entrepreneurs in Ethiopia and an opposing geographic country, Gambia. The first three studies looked at finances for women entrepreneurs beginning with Strobbe et al. (2015), who looked at how Ethiopia is falling behind its peers in the area of credit to the private sector and how it can be prevented from further happening. A study from Enterprise Surveys found that financial access was the most commonly reported issue across micro (41%), small (36%), and medium (29%) businesses in Ethiopia compared to a sub-Saharan Africa average of 24%, 20%, and 16% respectively (Strobbe et al., 2015). Singh et al. (2018) sought to identify issues related to women's entrepreneurship-pertinent to small-medium enterprises (SMEs) in Addis Ababa. The authors found problems in the areas of securing finances for establishing and running SMEs, lack of entrepreneurial and management competence and exposure, and problems in finding the markets and distribution networks (Singh et al., 2008).

Women entrepreneurs face a myriad of problems – and the women entrepreneurs in Egypt are no exception. Wube (2010) studied the factors that affected the performance of women entrepreneurs in MSEs. This article looked at business practices of women entrepreneurs in MSEs and the additional assistance they received from colleges. They found that individual characteristics of their enterprises affected their success (Wube, 2010). It also showed that lack of own premises (land), financial access, stiff competition, inadequate access to training, access to technology, and access to raw materials were the key economic factors that affected the performance of women entrepreneurs in MSEs (Wube, 2010). The study also found that conflicting gender roles, social acceptability, and networking with outsiders were the major social factors that affect these entrepreneurs (Wube, 2010). Furthermore, the main legal/administrative factors include access to policy makers, high amounts of tax and interest, bureaucracies and red tapes, and overall legal and regulatory environments (Wube, 2010). Conversely, Belwal et al., (2011), observed the effects of microfinance on women small-scale entrepreneurs in Addis Ababa, and found that women entrepreneurs who received microfinance experienced several issues (Belwal et al., 2011). While a study out of Gambia looked at the challenges faced by women entrepreneurs in the small enterprise sector (Della-Guista et al., 2006), the study attempted to discuss the gendered nature of the barriers to survival and growth, particularly the low productivity of sectors in which women entrepreneurs operate (Della-Guista et al., 2006). External factors such as cultural constraints played a role in women struggling to be successful in their business ventures. Women's roles are expected to be more traditional (Della-Guista et al., 2006).

Journeying to look at women entrepreneurs in the MENA (the Middle East and North Africa) region, we return to the familiar solid authentication of financial performance (Laffineur et al., 2018). The authors looked at women entrepreneurs in the MENA region where there exists a discriminatory environment for women entrepreneurs (Laffineur et al., 2018). They encountered difficulties in accessing finance, which constituted their principal obstacle to entrepreneurship (Laffineur et al., 2018). The experimenters felt that gendered stereotypes present large obstacles to entrepreneurship (Laffineur and Tavakoli, et. al, 2018). Furthermore, the experimenters hypothesize that the most important factor for these entrepreneurs was entrepreneurial education–learning how to acquire and expand their skills. Lastly, determining factors in the growth of these businesses was socioeconomic background.

For women entrepreneurs in Morocco, obstacles were identified in two main studies. El Marzougui, et al (2020) saw several obstacles to female entrepreneurship which are identified in specific areas of self-realization, independence, and personal growth (Bensassi et al., 2018). Financial issues keep propping up and shed light on the dynamic and creative economic strategies women entrepreneurs employ during a prolonged period of transit. The capacity to generate income seems to be severely restricted for women entrepreneurs. Lacking the legal entitlement to work in some circumstances, women entrepreneurs exhibit impressive creativity, by carving out an entrepreneurial niche in the informal sector. Cobblers, merchant women selling African products, chefs, and various others are often able to cover their daily costs consistently for a significant length of time. They are sometimes supported financially by associations that reinforce the solidarity within their community. Pickerill (2011) found this to be so. As Welsh et al. (2018) compared women's entrepreneurial firm performance to two measures (enrichment and interference) of business-family interface economic development in two emerging countries – Morocco and Turkey (Welsh et al., 2018) – Bouzekraoui and Driss Ferhane (Bouzekraoui & Ferhane, 2017) looked at women's entrepreneurship in Morocco through various indicators: the profile, the characteristics of the company created, and the main barriers behind the low rate of women's business creation in Morocco. The authors found that the majority of barriers encountered by women entrepreneurs are categorized into five major issues: (1) access to capital, (2) business performance, (3) networking, (4) training, and (5) work-family life balance (Bouzekraoui & Ferhane, 2017). These problems manifest across other relevant countries and similar groups of women entrepreneurs (Bouzekraoui & Ferhane, 2017).

What emerges from the literature review above is a clear picture of the myriad of challenges faced by the women entrepreneurs in Africa. The challenges face relating to a lack of access to initial and ongoing capital, lack of education in general and business education in particular, and a network structure that is adequate to support the goals the women entrepreneurs have. This network structure needs to be varied, have multiple levels of individuals and entities, have diverse levels of individuals and entities, and have individuals and entities from a wide cross section of society. What then of these challenges? How best can they be overcome so the women entrepreneurs in Africa can achieve their true potential? The question deserves an answer and it gets it in the last sections of this chapter. While women entrepreneurs in Africa face complex challenges and comprehensive problems they can be overcome, these challenges and problems are tied to sociocultural, political, and economic issues that will take a lot to unpack. Overcoming them requires an in-depth look at the opportunities that await.

9.4 Motivational Factors

The next set of research occurred in Egypt and looked at motivation factors. Specifically, self-determination theory (SDT) is used as a guiding framework for improved understanding of the motivation to form and develop a venture, with a special interest in motivation changes concerning the entrepreneurial ecosystem (Hailemariam, 2018). According to SDT, autonomous motivation and motivational change over time happen when all basic psychological needs for autonomy, competence, and relatedness are satisfied (Hailemariam, 2018). In contrast, Mulatu et al. (2018) looked at why women were motivated to become entrepreneurs in the first place, and what drives them to continue their progress (Mulatu et al., 2018). The results concluded that motivation for participating in entrepreneurial ventures by women may be classified into push and pull factors. Push factors are related to several factors, including frustration, job dissatisfaction, divorce, and boredom in their previous jobs. Pull factors included independence, autonomy, education, and family security (Mulatu et al., 2018). Hundera et al. (2019) rounded out this perspective, having researched the experience of role conflict and consequent coping strategies used by women entrepreneurs in sub-Saharan Africa (SSA) These latter authors found that sources of role conflict for women entrepreneurs in the context studied include family, business (work), social role expectations, and personal factors (Hundera et al., 2019).

Two studies looked at women entrepreneurs in the Middle East and North Africa, with the first finding the different challenges and motivations that women entrepreneurs, who adopt and use social media for marketing reasons, faced and experienced (Azoury et al., 2012). Women entrepreneurs in the Middle East and North Africa (MENA) regions are not provided with the required amount of education and resources when compared to the male entrepreneurs of those regions (Azoury et al., 2012). While Alghaith (2016) observed the different obstacles that women entrepreneurs, who utilize social media for marketing purposes, experienced women in the Middle East and North Africa (MENA) regions are not provided with the required amount of education and resources when compared to men in those regions (Alghaith, 2016). Gray et al. (2005) examined women entrepreneurs in the Kingdom of Morocco. The article began with a description of the negative stereotypes experienced by women in Islamic countries, briefly surveying Saudi Arabia, Iran, and Morocco as examples of differences in social practices in different Islamic countries (Gray et al., 2005). Self-sufficiency was seen as a major factor of motivation for the women entrepreneurs (Gray et al., 2005).

In a similar vein, Alene (2020) explored the determinants that influence women entrepreneurs' performance in micro and small enterprises in Gondar city, Northwest Ethiopia. Educational levels, previous entrepreneurial experiences, access to business training, access to finance, access to business information, government support, land ownership, and excessive taxes were seen as significant in explaining women entrepreneurs' performance on one hand Alene (2020). One area that is of paramount importance to women entrepreneurs is long-term survival. A study out of Ethiopia looked at the factors that influenced the long-term survival and viability of micro small and medium enterprises in Ethiopia (Bekele et al., 2008). Female-headed firms that ceased operation had an average lifetime of 3.2 years, while male-headed firms that ceased operation had an average lifetime of 3.9 years. Businesses that had failed were designated by challenges in obtaining loans from formal money lending organizations like commercial banks (61%), failure in investing profits (46%), bad managerial skills (54%), limited technical skills (49%), and low level of education (55%) (Bekele et al., 2008).

A more comprehensive study looked at some issues, namely: what are the factors motivating women to start their business ventures in South Africa, what are the factors that hinder women from growing their business like their male counterparts, and what are the barriers faced by women in terms of access to finance for entrepreneurial activities in South Africa (Witbooi & Ukpere, 2011)? Race was found to be a primary driver of financial access in South Africa, a gender gap that also existed and a gap that cuts across the races (Witbooi & Ukpere, 2011). The combination of race and gender disparities works largely to the detriment of Black women and women of color in these regions (Witbooi & Ukpere, 2011).

Benzing et al. (2009) interviewed 599 entrepreneurs in Kenya, Ghana, and Nigeria about how they started their businesses. The researchers found that a significant factor across the three countries represented in the study was the opportunity to increase income. Further analysis found three additional factors: family, external validation, and self-betterment (Benzing et al, 2009). While another article from

Kenya observed strategies by women in milk micro-enterprises. The study was carried out in Nairobi, Kiambu, and Nakuru districts. Strategies used by women to sustain their enterprises included accessing credit facilities from relatives and friends; using low-level technologies; operating their businesses illegally to avoid heavy licensing costs; and managing their enterprises close to home (Odero-Wanga et al., 2013).

The motivators for women entrepreneurs in Africa. On one level the women entrepreneurs in Africa want independence – especially from that which currently constrains them. The women entrepreneurs wanted to make a difference in the society in which they live; they wanted to contribute to the building of their countries' economies at different levels and in different ways, and equally important to move their families out of a lower-level income bracket to a higher standard of living – a feat that deserves prominent focus.

9.5 Experiences Across Gender Lines

Long ago studies on the Democratic Republic of Congo borrowed methodology from decades previously and evaluated the methodology of Oaxaca (1973) and Kautonen et al (2010) allowing us to explore the influence of gender on entrepreneurial performance and to bring out the differences in performance. The results suggest that female-managed and/or female-owned firms are less effective in terms of the number of employees than their male counterparts. More recently, a study looked at the influence of gender on entrepreneurial performance in the Democratic Republic of Congo (Janvier, 2012). Data from a survey conducted in May–June 2012 to SMEs (small-medium enterprises) completed by data from the 2006 and 2010 World Bank's Enterprise Surveys were used. Finally, relative to male entrepreneurs, female entrepreneurship positively influences the performance of firms in terms of growth of annual sales (Janvier, 2012).

An article by Stevenson et al. (2005) did a detailed assessment of the environment for the development of women's entrepreneurship in Cameroon, Mali, Nigeria, Rwanda, and Senegal which looked at gender-related implications of policies and programs to support women's economic development (WED) in the five countries and to ascertain whether there are similarities among these countries that could lead to the design of subregional or regional approaches toward creating a more conducive environment for WED (Stevenson et al., 2005). They found that for women to have equal opportunity to start and grow successful businesses, they must not only have equal access to education, training, skills, credit, information, business advisory services, business networks, markets, and property ownership as their male counterparts but be subject to the equitable administration of laws, regulations, and procedures about business, including such laws as those related to business entry, employment and labor, and property rights (Stevenson et al., 2005). Furthermore, when creating financial guidelines and programs designed to aid business, those involved in these creative processes should ensure that all genders are treated equally (Stevenson et al., 2005).

The analysis in a study from El-Hamidi (2011) was based on the 2003 MSE dataset made available by the ERF (Economic Research Forum of the Middle East and North Africa, Iran, and Turkey). The researchers had two objectives in this study: 1) To compare how women-owned MSEs differ from men-owned MSEs in human and financial capital; 2) To what degree do these differences/similarities affect the performance of the MSE. Using econometric techniques (i.e., logistical analysis), the sample size consisted of 5000 private MSEs. The paper evaluated the effect of aspects of human and financial capital on three measures of performance of MSEs in Egypt for men and women. Metrics for performance are increase in employment, rise in revenue, as well as growth in returns on sales (ROS) over a one-year period (El-Hamidi, 2011). Preliminary findings show gender differences in terms of general, and specific human capital, as well as financial capital, where women are at a disadvantage (El-Hamidi, 2011). Empirical findings reveal that women are better performers than men in generating revenues, even though their revenues are almost one-third of their males' counterparts (El-Hamidi, 2011). What's more interesting is that women were found to be no different from men in terms of employment growth or the efficiency of running their businesses (El-Hamidi, 2011).

A research project from Eswatini on women entrepreneurs explained that they started their firms with smaller start-up capital than men and are more likely to fund it from their sources, which reduces the size of their firm and sales level (Brixiova et al., 2019). However, women with higher educational levels started their firms with more capital than their less-educated counterparts (Brixiova et al., 2019). Moreover, women who received support from professional networks have higher initial capital, while those trained in financial literacy more often access external funding sources, including through their networks (Brixiova et al., 2019). The role of networks in the access of female entrepreneurs to start-up capital and firm performance in Eswatini, a country with one of the highest female unemployment rates in Africa, was found to be of critical importance (Brixiova et al., 2019). Latin America, sub-Saharan Africa, Eastern Europe, and Asia women entrepreneurs were looked at in a study by Sabarwal and Terrell (2009), and they found that women entrepreneurs had smaller scales of operation than their male counterparts, but gender performance gaps diminished when individual and environmental characteristics were taken into account.

Three studies on women entrepreneurs in Mauritius offer broad strokes to look into what is taking place regarding women entrepreneurs in that country. The first looked at the role of women entrepreneurs in the small island economy by focusing on the education sector and labor market access (Gokulsing, et al., 2014). Girls were found to outperform boys at all education levels, starting from primary, secondary, and tertiary levels, but their access to job opportunities was reduced (Gokulsing, et al., 2014). A second looked at the network composition of female entrepreneurs in Mauritius and also tried to identify the benefits derived from networking (Gungaphul et al., 2012). The study also identified the causes that hindered the networking practices of women entrepreneurs and found that female entrepreneurs tend to favor family and friends in their contact lists (Gungaphul et al., 2012). Their networks comprise more women than men (Gungaphul et al., 2012). A third looked at motivations, business models, and the socio-economic impacts of women business owners' enterprises (Rosunee et al., 2017). The concept of sustainability (people, planet, profit) underpinned the key aspects of their business, which proved that sustainability can be a positive driver of innovation and socio-economic change (Rosunee et al., 2017).

A comparison of women entrepreneurs in the Middle East and North Africa looked at women entrepreneurs in terms of entrepreneurial activity rates, entrepreneurial orientations, demographics, and their enterprise characteristics in the Middle East and North Africa (MENA) countries (Hala, 2012). The study suggests that the population of women in entrepreneurship has increased, however still smaller compareed to men in entrepreneurship (Hala, 2012).

Another study on women entrepreneurs in Tunisia compared perceptions of typical entrepreneurship support services across gender. The focus is on Tunisia, a developing country characterized by high levels of unemployment, particularly among women (Drine et al., 2010). Existing support services are inadequate for promoting female entrepreneurship and what solutions could fix this issue were also discussed (Drine et al., 2010). A fourth study examined the entrepreneurial intentions of young well-educated people in Tunisia, and accordingly, the attitudes and entrepreneurial intentions of females, and contrasting these with males, will provide an account of cultural attitudes toward the female enterprise. The researchers observed that cultural gender expectations continuted to play a key factor, however, it was most apparent in how women-led enterprises could be led, rather than in the attractiveness of entrepreneurship itself (Harbi et al., 2009). A final study on women entrepreneurs in Tunisia looked at the obstacles that are impeding women business creators to start their businesses. They found that the business creators' decision to launch their ventures is hindered mainly by the following factors: lack of managerial and business skills, poor training programs, and risk aversion. The researchers also wrote recommendations on ammeliorating obstacles facing future entrepreneurs and their respective projects (Sidrat et al., 2016). Going deeper into the continent showed that an important look needs to take place at the entrepreneurial motivations of women entrepreneurs in the United Arab Emirates (Tlaiss, 2015). It analyzed the impact of macro-social forces and cultural values on the motivation for entrepreneurship and found how the entrepreneurial motivations of Emirati women unfold in a complex interplay between pull and push motivational factors within the Arab patriarchal and Islamic contexts, thus lending credence to the post-materialism, legitimation, and dissatisfaction theories (Tlaiss, 2015).

What impact does gender have on entrepreneurship in Africa? When looking along gender lines, we see that the male entrepreneurs in Africa were more likely to be higher educated, have stronger network structures, and have more access to financial resources than their female counterparts. The male entrepreneurs were also more likely to be supported by political, sociocultural, and economic policies that showed a preference for male entrepreneurship over female entrepreneurs. The dominance of males in African countries speaks to the patriarchal society that often dictates these societies. A patriarchal nature permeates the educational, social, and political systems and policies and eventually bleeds into the economic systems through entrepreneurship and other areas.

9.6 Opportunities and Hope

A research study on women entrepreneurs in Algeria, Morocco, and Tunisia compared cultures, with data from these three traditional societies and two secular-rational societies, France and Belgium (Cheragi et al., 2014). This data was surveyed in the Global Entrepreneurship Monitor, randomly sampling 39,336 women, including 2306 entrepreneurs. Analyses show that women entrepreneurs have growth expectations based on their background and increased by their competence and opportunity-motive, which also promote business networks around their firms (Cheragi et al., 2014). The authors found that the formation of expectations is embedded in the networks in the way that networking increases expectations and is embedded in the culture in the way that, surprisingly, expectations are higher in traditional culture than in secular-rational culture (Cheragi et al., 2014).

Coughlin et al. (2002) examined subject groups of illiterate women entrepreneurs in Algeria to observe how they improve the economics of their country and found that feminism is a key determinant in encouraging women to become more involved in their country economically and in terms of development, however, it is unacknowledged by the general population (Halladay et al., 2003; Coughlin et al., 2002). A study from another country – Angola – looked at how women prosper in African countries as businesswomen (Siun et al., 2017). The authors noted that some women from that country have an unfair advantage of being born into an already successful family, so they use this and their subsequent higher educational levels to create advantages and create more wealth through their businesses (Siun et al., 2017).

What are the opportunities in entrepreneurship available to women entrepreneurs in Africa? The literature points to lucrative experiences as possible in the areas of the service industry and related industries. These opportunities also support an arena of thinking that sees such opportunities as a logical and steady source of income and source of survival for the women, their families, their communities, and the various countries' economies.

9.7 Success Factors

The relationship between elaborate and proactive planning and success was supported for the business' success factors size and expert evaluation, but not supported for growth in the South African study on women entrepreneurs (Frese et al., 2007).

However, the second set of researchers identified the success factors that motivate women entrepreneurs to contribute to economic growth and identify the barriers that impede the start-up, management, and growth of their business ventures (Akhalwaya & Havenga, 2012). They further found that developmental opportunities and networking as well as cultural and social values are major barriers hindering success (Akhalwaya & Havenga, 2012).

Women entrepreneurs perceive semiformal capital, such as loans from MFIs, SACCOS, ROSCAs, and VICOBA, as the most accessible external capital. Governmental subsidies are ranked second, followed by informal capital, such as loans from family, friends, and investors as financial sources for women entrepreneurs (Lindvert, et al., 2015). A further look regarding this took place in a study that looked at the growth of women entrepreneurs in Dar es Salaam city, Tanzania, and looked at strengths, weaknesses, opportunities, and threats. They found that the major strength was the need for financial independence, which is an attractive feature to women entrepreneurs (Benard et al., 2012). The major opportunity was the desire to be their boss, which directly leads to the world of entrepreneurship (Bernard et al., 2012). The weakness was most women entrepreneurs failed because of the lack of education, and the major threat was most women entrepreneurs had the pressure of child care in their families Autant-Bernard (2012).

In West Africa, researchers tested if a psychology-based personal initiative training approach that teaches and promotes a proactive mindset that focuses on entrepreneurial behaviors can have more success for women (Campos et al, 2017). The personal initiative mindset is key to entrepreneurial success since it involves looking for ways to differentiate oneself from other businesses, to anticipate problems, to better overcome setbacks...the "born with an entrepreneurial personality" versus "made by learning specific entrepreneurial practices viewpoints'" (Campos et al, 2017). By showing that training can teach people to develop a mindset with attributes like proactiveness that is often assumed to be innate, women entrepreneurs would be expected to be more successful (Campos et al, 2017).

Overall we see that for women entrepreneurs in Africa, their success hinges initially on proper training in overall educational fields and training in the particular area of business development and management. These previous factors form what are called human capital features. Other factors also play a prominent role, namely, rich network structures from family members, friends, and other business owners and government officials. These previous factors form what are called social capital features. Finally, the women entrepreneurs' success relies on access to financial resources at the start-up and during the continuation of the business. This last branch forms what is called financial capital features. It is these three forms of human capital, social capital, and financial capital that underlie the essence of what is needed for women entrepreneurs to be successful in Africa.

9.8 Statistics on Women on the African Continent

Table 9.1 depicts some key statistics for women in certain African countries. For the most part, the percentage of women in each of the countries listed below accounted for approximately 50 percent of that country's population, except for Malawi, where the percentage of women occupying that country stood at approximately 70 percent. The three countries with the largest percentage of women business owners are Uganda (38.2%), Ghana (37.2%), and Botswana (36%). For women's labor force participation rates by industries, the countries with the largest level of labor force participation were Nigeria, with 65.4% of women in the professional and technical industries, followed by Algeria with 44.2% of women in the professional and technical industries and then South Africa with 43.8% of the women in the informal sectors. Women in most countries were primarily in the middle-income sectors. In terms of women's educational levels, the countries that had the highest level of educated women were found to be Tunisia, with 68.9% of the women in that country completing university, and Botswana, with 59.1% of the women in that country completing university and Egypt with 42% of the women completing university.

This chapter presented a solid and extensive representation of women entrepreneurs in a wide variety of African countries. As the coverage traveled from country to country, a picture emerged that shows the key elements that impact women entrepreneurs across that continent. The key factors can be delegated to three key areas. The first is human capital, which refers specifically and generally to elements such as education, experience, skills, and knowledge and how these four variables can help the African women entrepreneurs with their businesses. The second is network capital, which looks at a matrix of connections for someone to others they know, who can help them with their entrepreneurial ventures. These connections include family members, friends, associates, other business owners, individuals that work at various institutions, and all who can help the African women business owners with their businesses. The third is the financial capital and primarily the lack of access that African women entrepreneurs often encounter, especially when compared to their male counterparts. What motivates the African women entrepreneurs to start their own business includes a sense of independence and the flexibility it provides for the African women entrepreneurs. Overall this chapter paints a very comprehensive picture of what it means to be and operate as a women entrepreneur in Africa.

In Tunisia, the long-term objective is for women entrepreneurs to be mainstreamed in all MSE policies and projects (i.e., for all MSE policies and projects to be gender aware), women – because of the particular constraints they continued to face and the discrimination that they have faced in the past – may also require specific "catch-up" interventions (Marcucci, 2014). This included interventions dealing with gender-specific constraints (e.g., lack of education and training or access to resources) that limit their participation in mainstream projects (Marcucci, 2014). In a similar vein, in a second study on women entrepreneurs in Tunisia, Mohamed et al., (2020) looked at the situation of female entrepreneurship in rural areas, and the various obstacles that help explain both the work patterns and the home conditioning

Table 9.1 Key statistics on women in African countries

Countries	Percent of women in that country	Percent of wbo in that country	Percent by industry	Incomes	Other statistical data – wbo (age, education, marital status, children)
Algeria	49.48	7.3	Professional & technical workers (44.2)	Upper middle	age:~ 18–44 Education: 25% completed university
Angola	50.53	30.3		Upper middle	Age: 15–24 Education: <0.7 attend university However 71% are literate
Botswana	51.64	36.0	Public sector (37%) – doctors, police, teachers, civil servants)	Upper middle	Education: 59.1% completed university Age: 18–24
Egypt	49.8	4.1	Professional & technical workers (38.6)	Lower middle	Age: ~18–34 Education: 42% completed university
Ethiopia	49.97	15.0	Professional & technical workers (32.9)	Low	Age: <49 Education: 5.2% completed university
Ghana	49.32	37.9	Professional & technical workers (31.8)	Lower middle	Age: 20+ Education: 13.53% tertiary
Malawi	70.69	31.1	Professional & technical workers (36.0)	Low	Age: 24+ Education: 2.9% completed university
Morocco	50.39	26	Labor force (23%) Professional and technical workers (35.65%)	Lower middle	Age: ~18–44 Education: 0% completed university
Nigeria	49.33	25.3	Professional & technical workers (65.4)	Lower middle	Age: 18–24 Education: 36% university
South Africa	50.72	21.5	Informal sector (43.8%)	Upper middle	Age: 15–24 Education: 1.3% tertiary completion (university)
Tunisia	50.42	10.9	Professional & technical workers (42.5)	Lower middle	Age: 28–40 Education: 68.9% from university
Uganda	50.74	38.2	Professional & technical workers (40.4)	Low	Age: 18–24 Education: 28.6% completed university

(continued)

Table 9.1 (continued)

Sources:
Algeria, Angola, Botswana, Egypt, Ethiopia, Ghana, Malawi, South Africa, Tunisia, Uganda, Nigeria
https://newsroom.mastercard.com/wp-content/uploads/2019/11/Mastercard-Index-of-Women-Entrepreneurs-2019.pdf
Botswana industry: http://documents1.worldbank.org/curated/en/290451565250128807/pdf/Crossovers-in-Botswana-Women-Entrepreneurs-Who-Operate-in-Male-Dominated-Sectors-Output-for-Women-Entrepreneurship-Study.pdf
Morocco
https://www.econstor.eu/bitstream/10419/180004/1/ofel-2018-p390-401.pdf (% wbo)
https://pdf.usaid.gov/pdf_docs/PA00X4H8.pdf (type of work force)
https://www.doingbusiness.org/content/dam/doingBusiness/country/m/morocco/MAR.pdf (income class type)
Egypt women population
https://countrymeters.info/en/Egypt
Everyone else's women population
https://tradingeconomics.com/
Education; Egypt, Algeria, Morocco
https://www-proquest-com.ezproxy.siena.edu/docview/1095717760/fulltextPDF/290474A98650413EPQ/1?accountid=13858
Education: Angola
https://borgenproject.org/education-and-literacy-in-angola/
Education: Tunisia https://www.infodev.org/infodev-files/resource/idi/document/Women%20Entrpreneurship%20in%20Tunisia,%20Rahman%20Soukaina.pdf
https://www.universityworldnews.com/post.php?story=20201201123459402#:~:text=Challenges%20facing%20Tunisian%20women&text=The%20percentage%20of%20female%20graduates,Higher%20Education%20and%20Scientific%20Research.
Uganda
https://www.sciencedirect.com/science/article/pii/S2590291120300127
Nigeria education
https://www.statista.com/statistics/1130794/number-of-university-students-in-nigeria/#:~:text=In%202017%2C%20Nigerian%20universities%20counted,slightly%20lower%20among%20bachelor%20students.
Botswana education
https://www.statsbots.org.bw/sites/default/files/Tertiary%20Education%20Statistics%202018.pdf
Botswana percent by industry
https://www.gov.bw/civil-registration/gender-disaggregated-data
Ethiopia education
https://borgenproject.org/top-10-facts-about-girls-education-in-ethiopia/#:~:text=Only%205.2%20percent%20of%20women,almost%2011%20percent%20of%20men.
Ethiopia age
https://cdn.atria.nl/epublications/IAV_B00085318.pdf
South Africa education
https://tradingeconomics.com/south-africa/ratio-of-female-to-male-tertiary-enrollment-percent-wb-data.html#:~:text=School%20enrollment%2C%20tertiary%20(gross),compiled%20from%20officially%20recognized%20sources.
Malawi education
https://borgenproject.org/girls-education-in-malawi-2/
Ghana education
https://borgenproject.org/top-10-facts-about-girls-education-in-ghana/
South Africa percent by industry
http://www.statssa.gov.za/?p=11375#:~:text=Women%20accounted%20for%2043%2C8,with%20men%20dominating%20the%20rest

model obstacles include personal aspects as well as economic and demographic elements (Mohamed, et al., 2020).

Kane (2013), meantime, examined the relationships between microcredit activities and the social capital of Senegalese female entrepreneurial loan recipients, especially the impact and success of microcredit as they are intimately tied to the female borrowers' social capital (Kane, 2013). In contrast, Pelligrini (2018) looked at the barriers that women entrepreneurs in Sierra Leone faced when it came to engaging in entrepreneurship. Entrepreneurship education (EE) was recognized as playing a significant role in an economic system and especially in a system catering to women entrepreneurs (Mbeteh et al , 2018). Thus, the last set of studies showed that education is said to be one of the factors that play a key role in the success of women entrepreneurs in Africa.

Employing grounded theory based on comparative case studies of informal microfinance businesses in East Africa, a theoretical foundation was built for studying the establishment and evolution of family firms in emerging markets by Khavul et al. (2009). The authors showed that East African entrepreneurs not only use both strong family and strong community ties to establish and grow businesses, but they also use strong community ties to counterbalance the obligations that strong extended family ties create (Khavul et al., 2009). In addition, it was shown that economic informality presents opportunities for some entrepreneurial businesses, but not others, to cycle rapidly from opportunity to opportunity as they maneuver toward higher value-creating ventures (Khavul et al., 2009).

Two other studies analyzed the informal sectors and the contributions to the existence and persistence of women entrepreneurs. One analyzed the characteristics of women entrepreneurs operating in the informal sector in Mauritius and investigated the impact of women's entrepreneurship on their livelihoods and that of their families informal sector, which has provided a self-employment outlet for unemployed and retrenched women in Mauritius (Kasseeah & Tandrayen-Ragoobur, 2014). Another research was whether women entrepreneurs operating in the informal sector in the small island economy of Mauritius have been affected by the global financial crisis (Kasseeah, et al., 2014). Findings suggest that they have been affected by revenue factors by earning less than they would have normally (Kasseeah, et al., 2014).

Thus, the informal sector, absent formal channels, and government avenues and support provide a rich gateway for women entrepreneurs to pursue an income earning potential in Africa.

Saffu et al. (2008) studied the strategic, firm-level factors related to the business performance of Ghanaian women. In an uncertain economic environment like Ghana's, where business owners are more concerned with manipulating scarce and limited resources, the authors felt that perhaps it was not wise to plan strategically but rather plan with a short-term focus (Saffu et al., 2008). Instead of a sophisticated strategic planning regime, the authors emphasized that a focus must be placed on short-term measures to tackle the continuing changes in the business environment (Saffu et al., 2008).

The chapter took us on a journey that looked comprehensively at the literature surrounding women entrepreneurs in 54 African countries and covered critical and life-altering issues such as motivators for the women entrepreneurs starting their businesses, the challenges the women entrepreneurs face, the problems they encounter, how the women entrepreneurs compared to their male counterparts, and finally the critical role they play financially, socially, politically, and environmentally to their families, communities, and countries' economies. This prominent importance of women entrepreneurs in these African countries does not always get the credit it deserves – nor is their contribution to innovation, and primary and secondary employment glorified but the contribution of women entrepreneurs in Africa cannot be negated or overlooked, their contribution is too pronounced and significant to the informal and the formal economic sectors and their families, communities, and countries.

References

Abdirahman, Z. Z., & Sauvée, L. (2016). Non-technological innovations from organisational design and change perspectives in VSEs and SMEs: The case of management systems. *International Journal of Organisational Design and Engineering, 4*(3–4), 177–194.

Adisa, T. A., Abdulraheem, I., & Mordi, C. (2014). The characteristics and challenges of small businesses in Africa: an exploratory study of nigerian small business owners. *Petroleum-Gas University of Ploiesti Bulletin, Technical Series, 66*(4).

Alene, E. T. (2020). Determinants that influence the performance of women entrepreneurs in micro and small enterprises in Ethiopia. *Journal of Innovation and Entrepreneurship, 9*, 1–20.

Autant-Bernard, C. (2012). Spatial econometrics of innovation: Recent contributions and research perspectives. *Spatial Economic Analysis, 7*(4), 403–419.

Azoury, N., & Kamel, S. (2014). Education in the Middle East: Challenges and opportunities. *Business and Education in the Middle East*, 99–130.

Bardasi, E., & Wodon, Q. (2006). *Measuring time poverty and analyzing its determinants: Concepts and application to Guinea*.

Bekele, E., & Worku, Z. (2008). Factors that affect the long-term survival of micro, small and medium enterprises in Ethiopia. *South African Journal of Economics, 76*(3), 548–568.

Belwal, R., & Teshome, H. (2011). Chat exports and the Ethiopian economy: Opportunities, dilemmas and constraints. *African Journal of Business Management, 5*(9), 3635.

Benzing, C., & Chu, H. M. (2009). A comparison of the motivations of small business owners in Africa. *Journal of Small Business and Enterprise Development, 16*(1), 60–77.

Brixiová, Z., & Kangoye, T. (2018). Skills as a barrier to women's start ups: A model with evidence from Eswatini. *ERSA Working Paper, 768*.

Brixiová, Z., & Kangoye, T. (2019). Networks, start-up capital and womens entrepreneurial performance in Africa: Evidence from Eswatini. In *High-growth Women's Entrepreneurship* (pp. 13–31). Edward Elgar Publishing.

Campos, F., Frese, M., Goldstein, M., Iacovone, L., Johnson, H. C., McKenzie, D., & Mensmann, M. (2017). Teaching personal initiative beats traditional training in boosting small business in West Africa. *Science, 357*(6357), 1287–1290.

Cheraghi, M., Setti, Z., & Schøtt, T. (2014). Growth-expectations among women entrepreneurs: Embedded in networks and culture in Algeria, Morocco, Tunisia and in Belgium and France. *International Journal of Entrepreneurship and Small Business, 23*(1–2), 191–212.

Della-Giusta, M., & Phillips, C. (2006). Women entrepreneurs in the Gambia: Challenges and opportunities. *Journal of International Development: The Journal of the Development Studies Association, 18*(8), 1051–1064.

Drine, I., & Grach, M. (2010). *WIDER Working Paper No. 2010/100 supporting women entrepreneurs in Tunisia.*

El Marzougui, M., Messaoudi, N., Dachry, W., Sarir, H., & Bensassi, B. (2020, November). *Demand driven MRP: Literature review and research issues.* In 13ème Conférence internationale de modélisation, optimisation et simulation (MOSIM2020), 12–14 Nov 2020, Agadir, Maroc.

Gray, K. R., & Finley-Hervey, J. (2005). Women and entrepreneurship in Morocco: Debunking stereotypes and discerning strategies. *The International Entrepreneurship and Management Journal, 1*(2), 203–217.

Gungaphul, M., & Kassean, H. (2012, July). *An insight into the networking approaches of women entrepreneurs in Mauritius.* In European business research conference proceedings.

Hailemariam, A. T., & Kroon, B. (2018). Redefining success beyond economic growth and wealth generation: The case of Ethiopia. In *Women entrepreneurs and the myth of 'underperformance'* (pp. 3–19). Edward Elgar Publishing.

Halliday, S. V., & Christy, R. (2003). Smaller businesses, marketing relationships and shared values. In *Academy of marketing SIG in marketing/entrepreneurship symposium.* University of Gloucestershire.

Hattab, H. (2012). Towards understanding female entrepreneurship in Middle Eastern and North African countries: A cross-country comparison of female entrepreneurship. *Education, Business and Society: Contemporary Middle Eastern Issues.*

Kautonen, T., Zolin, R., Kuckertz, A., & Viljamaa, A. (2010). Ties that blind? How strong ties affect small business owner-managers' perceived trustworthiness of their advisors. *Entrepreneurship and Regional Development, 22*(2), 189–209.

Lindvert, M., Yazdanfar, D., & Boter, H. (2015). Perceptions of financial sources among women entrepreneurs in Tanzania. *African Journal of Economic and Management Studies, 6*(2), 197–218.

Marcucci, P. N. (2014). *Jobs, gender and small enterprises in Africa and Asia: Lessons drawn from Bangladesh, the Philippines, Tunisia and Zimbabwe.*

Mbeteh, A., & Pellegrini, M. M. (2018). Entrepreneurship education in developing countries: A study of the key challenges in Sierra Leone. *African Entrepreneurship: Challenges and Opportunities for Doing Business*, 89–116.

Mboko, S., & Smith-Hunter, A. E. (2009). Women entrepreneurs in Zimbabwe: A case study. *Journal of Global Business Issues, 3*(1), 157.

Mohamed, M. E., Hewedi, M., Lehto, X., & Maayouf, M. (2020). Marketing local food and cuisine culture online: A case study of DMO's websites in Egypt. *International Journal of Tourism Cities, 6*(4), 1045–1068.

Mulatu, M., & Prasad, R. D. (2018). A study on factors affecting success of women entrepreneurs in Tigray region, Ethiopia. A case study of Adigrat town entrepreneurs. *American Journal of Economics and Business Management, 1*(1), 10–19.

Mwatsika, C. (2015). Entrepreneurship development and entrepreneurial orientation in rural areas in Malawi. *African Journal of Business Management, 9*(9), 425–436.

Oaxaca, R. (1973). Male-female wage differentials in urban labor markets. *International Economic Review*, 693–709.

Rosunee, S., & Peer, A. (2017). Innovative women entrepreneurs in the green economy: Two case studies from Mauritius. In *ICoSI 2014: Proceedings of the 2nd international conference on sustainable innovation* (pp. 169–178). Springer.

Sabarwal, S., & Terrell, K. (2009). *Does gender matter for firm performance? Evidence from Eastern Europe and Central Asia* (IZA Discussion Papers No. 3758).

Saffu, K., Obeng Apori, S., Elijah-Mensah, A., & Ahumatah, J. (2008). The contribution of human capital and resource-based view to small-and medium-sized tourism venture performance in Ghana. *International Journal of Emerging Markets, 3*(3), 268–284.

Singh, S. K., & Gaur, S. S. (2018). *Entrepreneurship and innovation management in emerging economies*. Management decision

Stevenson, L., & St-Onge, A. (2005). *Support for growth-oriented, women entrepreneurs in Ethiopia, Kenya, and Tanzania: An overview report.*

Strobbe, F., & Alibhai, S. (2015). *Financing women entrepreneurs in Ethiopia*. World Bank.

Tengeh, R. K., & Nkem, L. (2017). Sustaining immigrant entrepreneurship in South Africa: The role of informal financial associations. *Sustainability, 9*(8), 1396.

Welsh, D. H., Kaciak, E., & Shamah, R. (2018). Determinants of women entrepreneurs' firm performance in a hostile environment. *Journal of Business Research, 88*, 481–491.

Chapter 10
Religiosity and Business Owners

Owolabi Kuye, Ayodele Oniku, and Temitope A. Oje

10.1 Introduction

In a broader concept, the significance of a religion is to affect individual behavioral dispositions which happens through religion indoctrination, and application of tenets to behaviors and relationship with fellow man and daily activities. Relatedly, the effects of religion and religiosity on business owners in business management and operations manifest in many forms in terms of the types of business engaged in; the disposition to gains and profits; the types of customers and their faith or beliefs; the disposition to and operations of customer services; the types of goods or services offered; etc. The business owner's faith and depth of involvement in religious activities go a long way to determine business operations.

Religion is the faith confessed to by business owners and that upholds their beliefs in spiritual activities and decisions which may affect their lives and business decisions. The religiosity factor is explicated in the depth of application and involvement in religion tenets and doctrines. In Nigeria the three (3) popular religions are Christianity, Islam, and African traditional religion (ATR), and sizeable number of citizens, including business owners who put their faiths in religion of choice/preferences among the religions. The fact about Nigeria is that it is acclaimed to be a

O. Kuye (✉)
Department of Business Administration, Faculty of Management Sciences, Universityof Lagos, Akoka-Yaba, Nigeria
e-mail: okuye@unilag.edu.ng

A. Oniku
University of Lagos, Akoka-Yaba, Nigeria
e-mail: ooniku@unilag.edu.ng

T. A. Oje
Business Administration Department, University of Lagos, Akoka-Yaba, Nigeria
e-mail: temitope.oje@ubagroup.com

S. D. Carter (ed.), *COVID-19, Supply Chain, Climate Change, and Sustainable Development in Africa*, Springer Proceedings in Business and Economics,
https://doi.org/10.1007/978-3-031-26121-3_10

religious nation in that an average citizen is a devotee of one faith or the other, and this is characterized in the many features of business among small business owners, for instance, many a times the businesses' names, the industry, the location, etc. It isn't a strange thing to find out that larger percentage of butchers and cattle dealers irrespective of the region of the country are Muslims; larger percentage of pub or beer parlor owners are Christians; etc. Also, business name can reveal the faith of the owner, e.g., El Shaddai Enterprises, Al Amin Nig. Ltd., Good Shepherd & Sons Ltd., Insha Allah Motors Ent, etc.

The pertinent question remains: "does the depth and level of involvement of religious activities among business owners interpret to piousness, morality, and decency expected in business?" To what extent do business owners follow the tenets and doctrines of business in their respective operations? Does the Allport and Ross' (1967) intrinsicness and extrinsicness prevail in the business owners' personality, and what side of the religiosity scale prevail in business operation vis-à-vis business owners' personality?

10.2 Literature Review

10.2.1 Religiosity

According to McDaniel and Burnett (1990), religiosity is "the faith a person has in God" (Patwardhan et al., 2012, p. 62) which invariably influences lifestyle, beliefs, personality, sexuality, and general behavioral disposition to taking certain decision which might include business decision and entrepreneurs (Allport & Ross, 1967; Kavas et al., 2019). The pertinence of a strong belief of a Supreme Being among a larger percentage of the people of sub-Saharan Africa makes the case of religion and religiosity stronger and indisputable in the region. Largely, faiths and beliefs of people are found in the popular religions – Christianity, Islam, and traditional practice. Thus, Onaiyekan (2011) stresses that the demography of religion in Nigeria is rooted in the three main and popular religious practices in the country – the African traditional religion (ATR), Islam, and Christianity. Strategically, the influence and dictates of religion cannot be compromised, neither disregard in marketing practice due to its influence on buying decisions in terms of consumption patterns and composition among consumers. In the same vein, it affects business owners and entrepreneurs who approach business operations from the dictates of religion tenets and doctrines as stipulated by the Holy books, e.g., The Bible or Hadith *or* the fraternity rules among ATR. The effects of the religion injunctions on business owners' decisions are evident on critical issues like offering right product, engagement in bribery, avoidance of adulterated products, the resolution to keep industry ethics, adherence to religion-approved trades and products, religion disposition to profit, giving, credit sales, interest charges, artificial scarcity, abnormal profit, giving and taking bribery, etc. all these issues that are recurrent and debatable issues in business are squarely dealt with and explicated in many religions' tenets and doctrines. The holy books like Bible, Quran, and Hadith are very comprehensive on these matters. Thus, the disposition of business owners, and the level the owners put such into practice in business management, is a signal to the intrinsicness and extrinsicness level (Allport & Ross, 1967).

Technically, religiosity expression and practice are not limited to consumption decisions and patterns shown by consumers, but it equally dictates the market or business behaviors of business owners when it comes to certain operational decisions. Miller and Hoffmann's (1995) study reveals that religiosity influence individual's level of risk preference and aversion, and gender disposition to risk preference and aversion. According to Kavas et al. (2019), the issue of religiosity is very important in family business because it anchors germane issues in family business management, and adherence to the religious tenets proffer solutions to many issues in business and be a guide to issues like extravagance, envy, corrupt practices, and pride – *nazar* and *riba*. The work of Astrachan, Astrachan, Campopiano, and Bau (2020) emphasizes the issues of *Dharma* and *Karma* and the interpretation in business, and the way they guide business owners' operations; it is stated that while *Dharma* stresses adherence to religious beliefs and putting doctrines into business operations, *Karma* stresses the need for promoting social equity and justice. In a new dimension, Goel and Misra (2019) study religiosity in terms of attitude toward business ethics (ATBE), and the gender difference among business owners shows that there is no difference between gender in disposition to ethical adherence in business but childhood, family culture, and system contribute to ATBE and that intra-religiosity (intrinsicness) has more influence than inter-religiosity (extrinsicness) on ATBE.

10.2.1.1 Intrinsicness

Allport and Ross' (1967) study on religious orientation states that intrinsic orientation is found among people who "find their master motive in religion" and "they are, so far as possible, brought into harmony with the religious beliefs and prescriptions" (p. 434). In other words, intrinsic-oriented religious people are shaped, controlled, and influence by the religious tenets and doctrines, and the fact is that this is not limited to Christianity practice as shown in the study. By today's practice, intrinsic orientation can be found in other faiths like Islam and other faiths (Goel & Misra, 2019; Al-hajla, 2017; Abou-Yuossef et al., 2015). Genia (1993) states that intrinsically religious people are genuinely committed to their faith. In other words, intrinsic-oriented religious people are more inward changed and reformed with the religion's tenets and doctrines, and this involuntarily affects their outward actions in business, relationship, and attitudinal behaviors to issues around them. Thus, it is hypothesized that:

- H1. That business owners that conduct business in accordance to the teachings and tenets of their faith build a better business operations, management, and relationship.

10.2.1.2 Extrinsicness

Allport and Ross (1967) defines religion extrinsicness as "persons with this orientation are disposed to use religion for their own ends" (p. 434). Thus, extrinsicness factor, in contrast to intrinsic orientation, embraces religion as a mean to an end, and not an end. According to Neyrinck et al. (2010), extrinsic orientation is based on utilitarian approach that helps one to attain personal motives or self-centered ends. In other words, an extrinsic-oriented person sees religion as an instrument to be used to accomplish personal gains and fulfilment. By and large, extrinsic orientation allows individuals to uphold religion lightly with no deeper conviction to be embraced for inner conviction but to accomplish external pursuits. Extrinsic motivation allows business owners to compromise religion tenets and doctrines to satisfy selfish ends, and not to allow religion tenets to direct course of life and business decisions. This is buttressed by the findings of Patwardhan et al. (2012) that intrinsic religiosity is a significant predictor of measurement of consumer ethics scale, but extrinsic otherwise. Thus, it is hypothesized:

- H2. That business owners that conduct business in accordance to personal benefits and self ends above religious teachings build a better business operations, management, and relationship.

10.2.2 Religion in Nigeria

History had it that traditional religion; contemporarily called African traditional religion (ATR) is as old as Africa. Thus, long before the advent of Christianity and Islamic religions, African people worshipped a Supreme Being with local interpretations, beliefs, and meanings in different cultures and tribes (Uthman & Abbas, 2014; Onaiyekan, 2011; Kitause & Achinuke, 2013). The earlier existence of ATR helps the penetration and acceptance of Christianity and Islam easily as a substitution to existing traditional worship in the different communities based on superiority attached to Christianity and forceful action attached to Islam (Mbabuike, 1996).

The dimension of religion practice in the country shows that it is an integral part of living among the citizens, and equally a factor that determines and predicts other areas of life and livelihood. Mbabuike (1996) expresses that Africans "live in a world that intermingles continuously with spirits, the ancestors, and the gods. They are incurably superstitious, naturally explaining life, death, and the mysteries of the universe through the divine interventions and will. There is nothing that does not have an explanation, the inaudible, the unseen and all-natural/supernatural phenomena. Islam and Christianity found in Africa an already religiously grounded and spiritually fervent population that served as an easy prey for conversion" (p. 401). Invariably, the characteristics expressed above dominate the premises of religiosity among Nigerians. It is an understatement to say that religiosity has equally permeated the business transactions and relationship in the manner it has influenced other facets of life and livelihood among the citizens. Today, businesses' names are largely a function of religiosity or with a tone of religious expression that would easily make one to know the faith or the religion affiliation or denomination that a business owner belongs. It is not uncommon to see businesses' names like El Shaddai Enterprises Ltd.; Alpha and Omega Ltd.; Allah Dey Enterprises Ltd.; El Amin International Ltd.; etc. All these show the level of religion fervency that people have brought into commercial activity. Importantly, the religious emblems or stickers, or names on businesses in many instances, determine the category of clients that form patronage. In other words, in many instances the interplay of religiosity and people's lives determines what they consume, who they patronize, and where they shop – Christians look for another Christians for business patronage and a Moslem looks for another Moslem's business for patronage in certain cases. Thus, it is hypothesized:

- H3. That business owners that switch between intrinsic and extrinsic orientations in business decision will build a better business operations, management, and relationship.

10.3 Methodology

The study adopted quantitative study with employment of existing scale of *Psychometric Evaluation of the Allport-Ross I/E Scale*, with the addition of another set of scale developed by researchers to enable a robust study. Thus, the focus is to find the level of intrinsic and extrinsic motivations of business owners that affect the level of integration of religion doctrines and tenets in business operations and, also, to understand if it is possible and what stage or situation that will business owners compromise religiosity and religion beliefs, tenets, and doctrines to favour business decision that will advance performances.

The study adopts triangulation method whereby both qualitative and quantitative techniques were employed; the quantitative approach was based on the adaptation of the wok of Allport-Ross I/E scale (1993) that measured religious orientation on the platforms of intrinsicness and extrinsicness, i.e., religious orientation is a function of intrinsic and extrinsic motivations. The qualitative approach employed oral interview of selected business owners based on the interpretation of the Allport-Ross I/E scale to give robustness to the findings. Importantly, additional 11 items were developed and added to the Allport-Ross I/E scale to measure religion-business interplay in business. The focus of the study is to investigate the disposition of business owners in the categories of SMEs to interplay of religious or faith beliefs and adherence to religions' doctrines and tenets in business operations. Thus, four hundred and twenty (420) questionnaires were distributed to business owners in retailing, wholesaling, food/culinary, government contractors, fashion and textile designing, event management, grocery and agro-allied, transportation and car dealership, bookshop merchants, spare parts retailing, electronics dealership and IT-related retailing, etc. A total of 244 questionnaires were successfully completed and returned which amounts to 58% of the population. The information about the business owners is culled from online yellow page – V-Connect and the membership record of SMEDAN (Small and Medium Enterprises Development Agency) in Lagos State. The study was carried out in Lagos State. The instrument was measured on five-point Likert scale, and the Cronbach alpha test for the reliability of the instrument showed 0.72.

10.4 Hypothesis Testing

In effort to arrive at the accurate and reliable findings from the data analyzed above, the data was subjected to statistical test (chi-square test) using SPSS software Version 23.0 to determine if there is any statistical relationship between the variables in each of the hypotheses. It is necessary to make assumptions or guesses about the population parameter involved.

10.4.1 Decision Rule

At 95% level of confidence, if χ^2C exceeds the tabulated value χ^2t, the null hypothesis will be rejected, while the alternative hypothesis holds otherwise and vice versa. Also, decision as to the validity or otherwise of a test could be taken if the p-value is less than 5% level of significance, the null hypothesis (H_1) is rejected and the alternative hypothesis (H_1) accepted. However, if the p-value is greater than 5% level of significant, the null hypothesis is acceptable and alternative hypothesis rejected.

Hypothesis One

H_1: Business owners are intrinsically motivated in the relationship between faith and business

Chi-square test

	Observed N	Expected N	Residual
Strongly agree	14	81.3	−67.3
Agree	206	81.3	124.7
Indifferent	24	81.3	−57.3
Total	**244**		

Test statistics

Business owners are intrinsically motivated in the relationship between faith and business	
Chi-square	287.246[a]
Df	2
Asymp. Sig.	0.000

[a]0 cells (0.0%) have expected frequencies less than 5. The minimum expected cell frequency is 81.3

10.4.2 Interpretation of Result

The chi-square result in hypothesis one shows that $\chi^2C = 287.246$, while the table value at 2 (3–1) degree of freedom is 5.99 at confidence interval of $\alpha = 0.05$, the null hypothesis will be rejected if the computed value of the test statistics χ^2C exceeds the tabulated value χ^2t. Therefore, since $\chi^2C > \chi^2t$, the null hypothesis was therefore rejected, while the alternative hypothesis was accepted, which states that business owners are intrinsically motivated in the relationship between faith and business. Also, the test is significant since the p-value (0.000) is less than 5% level of significance.

Hypothesis Two

H_2: Business owners are extrinsically motivated in the relationship between faith and business.

Chi-square test

	Observed N	Expected N	Residual
Strongly agree	6	48.8	−42.8
Agree	93	48.8	44.2
Indifferent	126	48.8	77.2
Disagree	18	48.8	−30.8
Strongly disagree	1	48.8	−47.8
Total	**244**		

Test statistics	
Business owners are extrinsically motivated in the relationship between faith and business	
Chi-square	265.959[a]
Df	4
Asymp. Sig.	0.000

[a]0 cells (0.0%) have expected frequencies less than 5. The minimum expected cell frequency is 48.8

10.4.3 Interpretation of Result

The chi-square result in hypothesis one shows that $\chi^2C = 265.959$, while the table value at 4 (5–1) degree of freedom is 9.49 at confidence interval of $\alpha = 0.05$, the null hypothesis will be rejected if the computed value of the test statistics χ^2C exceeds the tabulated value χ^2t. Therefore, since $\chi^2C > \chi^2t$, the null hypothesis was therefore rejected, while the alternative hypothesis was accepted, which states that business owners are extrinsically motivated in the relationship between faith and business. Also, the test is significant since the p-value (0.000) is less than 5% level of significance.

Hypothesis Three

H_3: Business decision that advances performances compromise faith's tenets and stipulations.

Chi-square test

	Observed N	Expected N	Residual
Agree	22	61.0	−39.0
Indifferent	147	61.0	86.0
Disagree	73	61.0	12.0
Strongly disagree	2	61.0	−59.0
Total	**244**		

Test statistics	
Business decision that advances performances compromise faith's tenets and stipulations	
Chi-square	205.607[a]
Df	3
Asymp. Sig.	0.000

[a]0 cells (0.0%) have expected frequencies less than 5. The minimum expected cell frequency is 61.0

10.4.4 Interpretation of Result

The chi-square result in hypothesis one shows that $\chi^2C = 205.607$, while the table value at 3(4–1) degree of freedom is 7.81 at confidence interval of $\alpha = 0.05$, the null hypothesis will be rejected if the computed value of the test statistics χ^2C exceeds the tabulated value χ^2t. Therefore, since $\chi^2C > \chi^2t$, the null hypothesis was therefore rejected, while the alternative hypothesis was accepted, which states that business decision that advances performances compromise faith's tenets and stipulations. Also, the test is significant since the p-value (0.000) is less than 5% level of significance.

10.5 Discussion of the Findings and Conclusion

The triangulation approach provides platform for robust analysis of the study. The findings reveal that issue of religiosity and the interplay of religion and business decision among business owners have many faces depending on the prevailing circumstances in business, personal lifestyle, morality decision, and decision in the choice between business and morality; thus, business owners have many faces in exhibiting religiosity. The practice of intrinsic religious orientation (Genia, 1993; Neyrinck et al., 2010) occurs outside business decision, and it is negative to business decision especially where issues of making extra profit; exaggeration of product features; and deception of consumers in order to make profit decisions are involved. In all these cases, business owners might not bring religion values and morality into practice to influence business decision. In other words, intrinsic religious orientation comes into practice on issue of personal piousness and decision on other nonbusiness/money in life. However, business owners do seek the face of God, Allah, or the Supreme Being when businesses experience downturn or poor patronage.

By and large, business owners' disposition to religiosity reveals tripartite orientation whereby intrinsic motivation has a role in the personal life of owners that can be separated from the commercial business, and this may be to accomplish certain celestial pursuits. And the extrinsic motivation comes into business owners' operations to achieve different goals outside religiosity or terrestrial pursuits, and this may further and deeply extend to prevalence of compromise of religion tenets and doctrines (that stress morality and ethics) to accomplish business performances. Also, the business-religion motivation focuses more on advancement of business performance; hence, the motivation is defined that business owners compromise religion tenets and doctrines, when necessary, to advance business performance and they as well religiosity and religion exercise to advance business lots when they believe the need arise as a spiritual exercise.

References

Abou-Yuossef, M., Kortam, W., Abou-Aish, E., & El-Bassiouny, N. (2015). Effects of religiosity on consumer attitudes towards Islamic banking in Egypt. *International Journal of Bank Marketing, 33*(6), 786–807.

Allport, G. W., & Ross, J. M. (1967). Personal religious orientation and prejudice. *Journal of Personality and Social Psychology, 5*(4), 432–443.

Al-Haijla, A. (2017). The Factors Influencing the Adoption of Wireless Application Protocols Banking in Saudi Arabia. *European Journal of Business Management, 9*(18), 60–68.

Astrachan, J.H., Astrachan, C.B., Campopiano, G., and Bau, M. (2020). values, Spirituality and Religion: Family Business and the Roots of Sustainable Ethical Behaviours. *Journal of Business Ethics, 163*(4), pp. 637–645.

Genia, V. (1993). A psychometric evaluation of the Allport-Ross I/E scales in a religiously Heterogenous sample. *Journal for the Scientific Study of Religion, 32*(3), 284–290.

Goel, Y. and Mishra, S. (2019). Cultural Understanding of Fathering and Fatherhood in India: An Exploration of Lived Experience. Consumer and Psychology, https://doi.org/10.1177/1354067X231154006

Kavas, M., Jarzabkowski, P., & Nigam, A. (2019). Islamic family business: The constitutive role of religion in business. *Journal of Business Ethics, 163*(4), 689–700.

Kitause, R. H., & Achinuke, H. C. (2013). Religion in Nigeria from 1900–2013. *Religion, 3*(18), 45–57.

Mbabuike, M. C. (1996). Skimming the new waves: A survey of new age religions in Nigeria. *Journal of Black Studies, 26*(4), 401–413.

McDaniel, S.W. and Burnett, J.J. (1990) Consumer Religiosity and Retail Store Evaluative Criteria. *Journal of the Academy of Marketing Science, 18*, 101–112.

Miller, A. S., & Hoffmann, J. P. (1995). Risk and religion: Explanation of gender differences in religiosity. *Journal for the Scientific Study of Religion, 34*(1), 63–75.

Neyrinck, B., Lens, W., Vansteenkiste, M., & Soenens, B. (2010). Updating Allport's and Batson's framework of religious orientations: A revaluation from the perspective of self-determination theory and Wulff's social cognitive model. *Journal for the Scientific Study of Religion, 49*(3), 425–438.

Onaiyekan, J. O. (2011). *Christianity: A foreign religion?* Gaudium Et Spes Institute.

Patwardhan, A. M., Keith, M. E., & Vitell, S. J. (2012). Religiosity, attitude towards business, and ethical beliefs: Hispanic consumers in the United States. *Journal of Business Ethics, 110*(1), 61–70.

Uthman, I. O., & Abbas, L. O. (2014). Muslim/Christian politics of religion in Nigeria: The Shari'ah application and the religious foundations of global Muslim engagement with modernity. *International Journal of Sociology and Anthropology, 16*(5), 169–179.